MANAGEMENT OF ORGANIC WASTE

Edited by **Sunil Kumar**
and **Ajay Bharti**

Management of Organic Waste

Edited by Sunil Kumar and Ajay Bharti

Published by InTech

Janeza Trdine 9, 51000 Rijeka, Croatia

Publishing Process Manager Maja Bozicevic

Technical Editor Teodora Smiljanic

Cover Designer InTech Design Team

Additional hard copies can be obtained from orders@intechopen.com

Management of Organic Waste, Edited by Sunil Kumar and Ajay Bharti

p. cm.

ISBN 978-953-307-925-7

For EU product safety concerns:
IN TECH d.o.o., Prolaz Marije Krucifikse Kozulić 3, 51000 Rijeka, Croatia,
info@intechopen.com or visit our website at intechopen.com.

Meet the editors

Er Sunil Kumar works in Environmental Engineering and Management, specializing in Solid and Hazardous Waste Management. He has been working at NEERI for the last 11 years. He is an Editorial Board member of Environmental Monitoring and Assessment, Open Waste Management Journal and also Regional Editor for Asian Countries, International Journal of Process Wastes Treatment and Serial Publications in New Delhi. He also edited special issues on various aspects of solid waste management for journals such as Environmental Monitoring and Assessment Journal of Air and Waste Management Association, International Journal of Environment and Pollution, International Journal of Environment and Waste Management, International Journal of Environmental Technology and Management, etc. He is one of the potential reviewers of Many International Journals including Environmental Science & Technology, Bio-resource Technology, Waste Management, Waste Management World, Chemical Engineering Journal, ASCE, Journal of Hazardous Materials, Ecological Engineering, Atmospheric Environment, Environmental Modeling and Software, Environmental Monitoring and Assessment, and Chemosphere. He has published 35 research papers in reputed international journals. Sunil Kumar has done research all over the world.

Dr Ajay Bharti is an Assistant Professor at NERIST, India. He has been working in the area of Environmental Science and Engineering for quite some time. His research area includes municipal solid waste management with an emphasis on processing aspects. He has published many well received papers in international journals.

Contents

Preface

Solid waste management is one of the important disciplines of environmental management. It is divided into two parts, dealing with biodegradable and non-biodegradable waste. The segregation of waste in most developing countries is a difficult task. This problem has a wide range of causes, including the lack of public knowledge of the problem. Lack of funds plays a small but very vital role.

Solid waste management is a wide and diversified field. Within this field, organic waste attracts a great deal of attention because of its chemical constituents. However, the few narrowly specialized resources on this type of waste are insufficient to reveal the complete chemistry of it. The book titled "Management of Organic Waste " is designed to provide a fundamental knowledge of the principles related to the management of organic waste. The chapters of the book are arranged logically and they offer an up-to-date approach to offer a better understanding of the chemistry used to treat organic waste as a raw material which results in a useful product. The breadth and depth of the material presented in this book will help to understand the different processing and disposal aspects of organic waste. The comparative aspects of processing and disposal, reflect the unique identity of the book. Lastly, each chapter with different sub-headings contains very good resources, and very clear concepts.

This publication will be extremely helpful to students, researchers, scientists, policy makers, and local waste management authorities.

Er Sunil Kumar, Scientist
Council of Scientific and Industrial Research-National Environmental Engineering Research Institute (CSIR-NEERI), Kolkata Zonal Laboratory
Kolkata, West Bengal,
India

Dr Ajay Bharti, Assistant Professor
North Eastern Regional Institute of Science & Technology [NERIST]
Nirjuli, Itanagar
Arunachal Pradesh
India

Part 1

Biogas from Organic Waste

1

Anaerobic Treatment and Biogas Production from Organic Waste

Gregor D. Zupančič and Viktor Grilc
Institute for Environmental Protection and Sensors
Slovenia

1. Introduction

Organic wastes under consideration are of natural origin that possess biochemical characteristics ensuring rapid microbial decomposition at relatively normal operating conditions. When considering the organic waste treatment we have generally in mind organic mineralization, biological stabilisation and detoxification of pollutants. Most common organic wastes contain compounds that are mainly well biodegradable. They can be readily mineralized either through biological treatment (aerobic or anaerobic), or thermo chemical treatment such as incineration, pyrolysis and gasification. The latter will not be treated in this work. Most organic wastes produced today originate in municipal, industrial and agricultural sector. Municipal waste (as well as municipal wastewater sludge) is generated in human biological and social activities and contains a large portion of organic waste readily available for treatment. Agricultural waste is common in livestock and food production and can be utilised for biogas production and therefore contribute to more sustainable practice in agriculture. Industrial wastes arise in many varieties and are the most difficult for biological treatment, depending of its origin. Namely, many industries use chemicals in their production in order to achieve their product quality and some of these chemicals are present in the waste stream, which is consequently difficult to treat. Recently, organic waste treatment has had a lot of attention, due to possibilities of energy recovery from these wastes as well as to prevent their adverse environmental effects. Energy recovery is possible through controlled release of chemically bound energy of organic compounds in waste and can be retrieved through chemical and biochemical processes. Most of the organic wastes appear in solid form; however contain up to 90% of moisture, therefore thermo-chemical treatment such as incineration cannot be applied. To address sustainability in the treatment of organic wastes, environmental aspect, energy aspect and economical aspect of the treatment processes should be considered.

Biodegradable organic waste can be treated with or without air access. Aerobic process is composting and anaerobic process is called digestion. Composting is a simple, fast, robust and relatively cheap process producing compost and CO_2 (Chiumenti et al. 2005, Diaz et al. 2007). Digestion is more sophisticated, slow and relatively sensitive process, applicable for selected input materials (Polprasert, 2007). In recent years anaerobic digestion has become a prevailing choice for sustainable organic waste treatment all over the world. It is well suited for various wet biodegradable organic wastes of high water content (over 80%), yielding methane rich biogas for renewable energy production and use.

Table 1 shows typical solid and organic substance contents and biogas yields for most frequent organic wastes, treated with anaerobic digestion.

Organic waste	TS^1 [%]	VS^2 in TS [%]	Biogas yield (SPB) [m^3kg^{-1} of VS]
Municipal organic waste	15-30	80-95	0.5-0.8
Municipal wastewater sludge	3-5	75-85	0,3-0,5
Brewery spent grain	20-26	80-95	0.5-1.1
Yeast	10-18	90-95	0.5-0.7
Fermentation residues	4-8	90-98	0.4-0.7
Fruit slurry (juice production)	4-10	92-98	0.5-0.8
Pig stomach content	12-15	80-84	0.3-0.4
Rumen content (untreated)	12-16	85-88	0.3-0.6
Vegetable wastes	5-20	76-90	0.3-0.4
Fresh greens	12-42	90-97	0.4-0.8
Grass cuttings (from lawns)	20-37	86-93	0.7-0.8
Grass silage	21-40	87-93	0.6-0.8
Corn silage	20-40	94-97	0.6-0.7
Straw from cereals	~86	89-94	0.2-0.5
Cattle manure (liquid)	6-11	68-85	0.1-0.8
Cattle excreta	25-30	75-85	0.6-0.8
Pig manure (liquid)	2-13	77-85	0.3-0.8
Pig excreta	20-25	75-80	0.2-0.5
Chicken excreta	10-29	67-77	0.3-0.8
Sheep excreta	18-25	80-85	0.3-0.4
Horse excreta	25-30	70-80	0.4-0.6
Waste milk	~8	90-92	0.6-0.7
Whey	4-6	80-92	0.5-0.9

[1]TS - total solids

[2]VS - volatile (organic) solids

Table 1. Types of organic wastes and their biogas yield

2. Basics of anaerobic digestion

This section deals with anaerobic waste treatment methods only, as the most advanced and sustainable organic waste treatment method. Anaerobic digestion (WRAP 2010) is *"a process of controlled decomposition of biodegradable materials under managed conditions where free oxygen is absent, at temperatures suitable for naturally occurring mesophilic or thermophilic anaerobic and facultative bacteria and archaea species, that convert the inputs to biogas and whole digestate"*. It is widely used to treat separately collected biodegradable organic wastes and wastewater sludge, because it reduces volume and mass of the input material with biogas (mostly a mixture of methane and CO_2 with trace gases such as H_2S, NH_3 and H_2) as by-product.

Thus, anaerobic digestion is a renewable energy source in an integrated waste management system. Also, the nutrient-rich solids left after digestion can be used as a fertilizer.

2.1 Biochemical reactions in anaerobic digestion

There are four key biological and chemical stages of anaerobic digestion:

1. Hydrolysis
2. Acidogenesis
3. Acetogenesis
4. Methanogenesis.

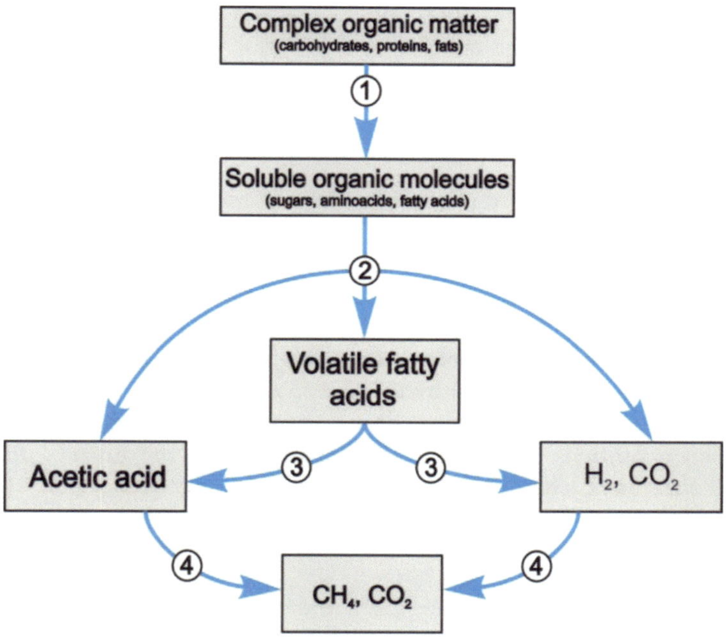

Fig. 1. Anaerobic pathway of complex organic matter degradation

In most cases biomass is made up of large organic compounds. In order for the microorganisms in anaerobic digesters to access the chemical energy potential of the organic material, the organic matter macromolecular chains must first be broken down into their smaller constituent parts. These constituent parts or monomers such as sugars are readily available to microorganisms for further processing. The process of breaking these chains and dissolving the smaller molecules into solution is called hydrolysis. Therefore hydrolysis of high molecular weight molecules is the necessary first step in anaerobic digestion. It may be enhanced by mechanical, thermal or chemical pretreatment of the waste. Hydrolysis step can be merely biological (using hydrolytic microorganisms) or combined: bio-chemical (using extracellular enzymes), chemical (using catalytic reactions) as well as physical (using thermal energy and pressure) in nature.

Acetates and hydrogen produced in the first stages can be used directly by methanogens. Other molecules such as volatile fatty acids (VFA's) with a chain length that is greater than acetate must first be catabolised into compounds that can be directly utilised by methanogens. The biological process of acidogenesis is where there is further breakdown of the remaining components by acidogenic (fermentative) bacteria. Here VFA's are generated along with ammonia, carbon dioxide and hydrogen sulphide as well as other by-products.

The third stage anaerobic digestion is acetogenesis. Here simple molecules created through the acidogenesis phase are further digested by acetogens to produce largely acetic acid (or its salts) as well as carbon dioxide and hydrogen.

The final stage of anaerobic digestion is the biological process of methanogenesis. Here methanogenic archaea utilise the intermediate products of the preceding stages and convert them into methane, carbon dioxide and water. It is these components that makes up the majority of the biogas released from the system. Methanogenesis is – beside other factors - sensitive to both high and low pH values and performs well between pH 6.5 and pH 8. The remaining, non-digestible organic and mineral material, which the microbes cannot feed upon, along with any dead bacterial residues constitutes the solid digestate.

2.2 Factors that affect anaerobic digestion

As with all biological processes the optimum environmental conditions are essential for successful operation of anaerobic digestion (Table 2). The microbial metabolism processes depend on many parameters; therefore these parameters must be considered and carefully controlled in practice. Furthermore, the environmental requirements of acidogenic bacteria differ from requirements of methanogenic archaea. Provided that all steps of the degradation process have to take place in one single reactor (one-stage process) usually methanogenic archaea requirements must be considered with priority. Namely, these organisms have much longer regeneration time, much slower growth and are more sensitive to environmental conditions then other bacteria present in the mixed culture (Table 3). However, there are some exceptions to the case:

Parameter	Hydrolysis/ Acidogenesis	Methanogenesis
Temperature	25-35°C	Mesophilic: 30-40°C Thermophilic: 50-60°C
pH Value	5.2-6.3	6.7-7.5
C:N ratio	10-45	20-30
Redox potential	+400 to -300 mV	Less than -250 mV
C:N:P:S ratio	500:15:5:3	600:15:5:3
Trace elements	No special requirements	Essential: Ni, Co, Mo, Se

Table 2. Environmental requirements (Deublein and Steinhauser 2008)

- With cellulose containing substrates (which are slowly degradable) the hydrolysis stage is the limiting one and needs prior attention.
- With protein rich substrates the pH optimum is equal in all anaerobic process stages therefore a single digester is sufficient for good performance.
- With fat rich substrates, the hydrolysis rate is increasing with better emulsification, so that acetogenesis is limiting. Therefore a thermophilic process is advised.

In aspiration to provide optimum conditions for each group of microorganisms, a two-stage process of waste degradation has been developed, containing a separate reactor for each stage. The first stage is for hydrolysis/acidification and the second for acetogenesis/methanogenesis. The process will be discussed in detail in section 3.

Microorganisms	Time of regeneration
Acidogenic bacteria	Less than 36 hours
Acetogenic bacteria	80-90 hours
Methanogenic archaea	5-16 days
Aerobic microorganisms	1-5 hours

Table 3. Regeneration time of microorganisms

2.2.1 Temperature

Anaerobic digestion can operate in a wide range of temperature, between 5°C and 65°C. Generally there are three widely known and established temperature ranges of operation: psychrophilic (15-20°C), mesophilic (30-40°C) and thermophilic (50-60°C). With increasing temperature the reaction rate of anaerobic digestion strongly increases. For instance, with ideal substrate thermophilic digestion can be approx. 4 times faster than mesophilic. However using real waste substrates, there are other inhibitory factors that influence digestion, that make thermophilic digestion only approx. 2 times faster than mesophilic.

The important thing is, when selecting the temperature range, it should be kept constant as much as possible. In thermophilic range (50-60°C) fluctuations as low as ±2°C can result in 30% less biogas production (Zupančič and Jemec 2010). Therefore it is advised that temperature fluctuations in thermophilic range should be no more than ±1°C. In mesophilic range the microorganisms are less sensitive; therefore fluctuations of ±3°C can be tolerated.

For each range of digestion temperature there are certain groups of microorganisms present that can flourish in these temperature ranges. In the temperature ranges between the three established temperature ranges the conditions for each of the microorganisms group are less favourable. In these ranges anaerobic digestion can operate, however much less efficient. For example, mesophilic microorganisms can operate up to 47°C, thermophilic microorganisms can already operate as low as 45°C. However the rate of reaction is low and it may happen that the two groups of microorganisms may exclude each other and compete in the overlapping range. This results in poor efficiency of the process, therefore these temperatures are rarely applied.

2.2.2 Redox potential

In the anaerobic digester, low redox potential is necessary. Methanogenic archaea need redox potential between -300 and -330 mV for the optimum performance. Redox potential can increase up to 0 mV in the digester; however it should be kept in the optimum range. To achieve that, no oxidizing agents should be added to the digester, such as oxygen, nitrate, nitrite or sulphate.

2.2.3 C:N ratio and ammonium inhibition

In microorganism biomass the mass ratio of C:N:P:S is approx. 100:10:1:1. The ideal substrate C:N ratio is then 20-30:1 and C:P ratio 150-200:1. The C:N ratio higher than 30 causes slower microorganisms multiplication due to low protein formation and thus low energy and structural material metabolism of microorganisms. Consequently lower substrate degradation efficiency is observed. On the other hand, the C:N ratio as low as 3:1 can result in successful digestion. However, when such low C:N ratios and nitrogen rich substrates are applied (that is often the case using animal farm waste) a possible ammonium inhibition must be considered. Ammonium although it represents an ideal form of nitrogen for microorganisms cells growth, is toxic to mesophilic methanogenic microorganisms at concentrations over 3000 mgL^{-1} and pH over 7.4. With increasing pH the toxicity of ammonium increases (Fig. 2).

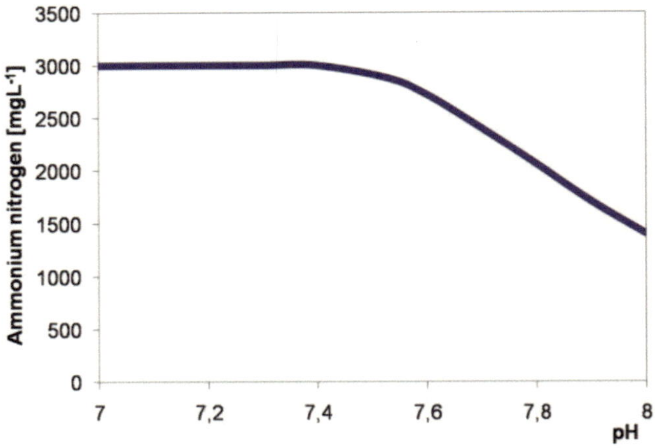

Fig. 2. Ammonium nitrogen toxicity concentration to methanogenic microorganisms

Thermophilic methanogenic microorganisms are generally more sensitive to ammonium concentration. Inhibition can occur already at 2200 mgL^{-1} of ammonium nitrogen. However the ammonium inhibition can very much depend on the substrate type. A study of ammonium inhibition in thermophilic digestion shows an inhibiting concentration to be over 4900 mgL^{-1} when using non-fat waste milk as substrate (Sung and Liu 2003).

Ammonium inhibition can likely occur when digester leachate (or water from dewatering the digested substrate) is re-circulated to dilute the solid substrate for anaerobic digestion. Such re-circulation must be handled with care and examined for potential traps such as ammonium or other inhibitory ions build up.

To resolve ammonia inhibition when using farm waste in anaerobic digestion several methods can be used:

- First possibility is carefully combining different substrates to create a mixture with lower nitrogen content. Usually some plant biomass (such as silage) is added to liquid farm waste in such case.
- Second possibility is diluting the substrate to such extent, that concentration in the anaerobic digester does not exceed the toxicity concentration. This method must be handled with care. Only in some cases dilution may be a solution. If the substrate requires too much dilution, a microorganisms washout may occur, which results in process failure. Usually there is only a narrow margin of operation, original substrate causes ammonium inhibition, when diluted to the extent necessary to stop ammonia inhibition, and already a washout due to dilution occurs.
- It is also possible to remove ammonium from the digester liquid. This method is usually most cost effective but rarely used. One of such processes is stripping ammonia from the liquid. It is also commercially available (GNS 2009).

2.2.4 pH

In anaerobic digestion the pH is most affecting the methanogenic stage of the process. pH optimum for the methanogenic microorganisms is between 6.5 and 7.5. If the pH decreases below 6.5, more acids are produced and that leads to imminent process failure. In real digester systems with suspended biomass and substrate containing suspended solids, normal pH of operation is between 7.3 and 7.5. When pH decreases to 6.9 already serious actions to stop process failure must be taken. When using UASB flow through systems (or other systems with granule like microorganisms), which utilize liquid substrates with low suspended solids concentration normal pH of operation is 6.9 to 7.1. In such cases pH limit of successful operation is 6.7.

In normally operated digesters there are two buffering systems which ensure that pH persists in the desirable range:

- Carbon dioxide - hydrogen carbonate - carbonate buffering system. During digestion CO_2 is continuously produced and release into gaseous phase. When pH value decreases, CO_2 is dissolved in the reactor solution as uncharged molecules. With increasing pH value dissolved CO_2 form carbonic acid which ionizes and releases hydrogen ions. At pH=4 all CO_2 is in form of molecules, at pH=13 all CO_2 is dissolved as carbonate. The centre point around which pH value swings with this system is at pH=6.5. With concentrations between 2500 and 5000 mgL^{-1} hydrogen carbonate gives strong buffering.
- Ammonia - ammonium buffering system. With decreasing pH value, ammonium ions are formed with releasing of hydroxyl ions. With increasing pH value more free ammonia molecules are formed. The centre point around which pH value swings with this system is at pH=10.

Both buffering systems can be overloaded by the feed of rapidly acidifying (quickly degradable) organic matter, by toxic substances, by decrease of temperature or by a too high loading rate to the reactor. In such case a pH decrease is observable, combined with CO_2 increase in the biogas. Measures to correct the excessive acidification and prevent the process failure are following:

- Stop the reactor substrate supply for the time to methanogenic archaea can process the acids. When the pH decreases to the limit of successful operation no substrate supply should be added until pH is in the normal range of operation or preferably in the upper portion of normal range of operation. In suspended biomass reactors this pH value is 7.4 in granule microorganisms systems this pH value is 7.0.
- If procedure from the point above has to be repeated many times, the system is obviously overloaded and the substrate supply has to be diminished by increasing the residence time of the substrate.
- Increase the buffering potential of the substrate. Addition of certain substrates which some contain alkaline substances to the substrate the buffering capacity of the system can be increased.
- Addition of the neutralizing substances. Typical are slaked lime ($Ca(OH)_2$), sodium carbonate (Na_2CO_3) or sodium hydrogen carbonate ($NaHCO_3$), and in some cases sodium hydroxide ($NaOH$). However, with sodium substances most precaution must be practiced, because sodium inhibition can occur with excessive use.

2.2.5 Inhibitory substances

In anaerobic digestion systems a characteristic phenomenon can be observed. Some substances which are necessary for microbial growth in small concentrations inhibit the digestion at higher concentrations. Similar effect can have high concentration of total volatile fatty acids (tVFA's). Although, they represent the very substrate that methanogenic archaea feed upon the concentrations over 10,000 mgL^{-1} may have an inhibitory effect on digestion (Mrafkova et al., 2003; Ye et al., 2008).

Inorganic salts can significantly affect anaerobic digestion. Table 4 shows the optimal and inhibitory concentrations of metal ions from inorganic salts.

	Optimal concentration [mgL^{-1}]	Moderate inhibition [mgL^{-1}]	Inhibition [mgL^{-1}]
Sodium	100-200	3500-5500	16000
Potassium	200-400	2500-4500	12000
Calcium	100-200	2500-4500	8000
Magnesium	75-150	1000-1500	3000

Table 4. Optimal and Inhibitory concentrations of ions from inorganic salts

In real operating systems it is unlikely that inhibitory concentrations of inorganic salts metals would occur, mostly because in such high concentrations insoluble salts would precipitate in alkaline conditions, especially if H_2S is present. The most real threat in this case is sodium inhibition of anaerobic digestion. This can occur in cases where substrates are wastes with extremely high salt contents (some food wastes, tannery wastes...) or when excessive use of sodium substances were used in neutralization of the substrate or the digester liquid. Study done by Feijoo et al. (1995) shows that concentrations of 3000 mgL^{-1} may already cause sodium inhibition. However, anaerobic digestion can operate up to concentrations as high as 16,000 mgL^{-1} of sodium, which is close to saline concentration of seawater. Measures to correct the sodium inhibition are simple. The high salt substrates must be pre-treated to remove the salts (mostly washing). The use of sodium substances as neutralizing agents can be substituted with other alkaline substances (such as lime).

Heavy metals also do have stimulating effects on anaerobic digestion in low concentrations, however higher concentrations can be toxic. In particular lead, cadmium, copper, zinc, nickel and chromium can cause disturbances in anaerobic digestion process. In farm wastes, e.g. in pig slurry, especially zinc is present, originating from pig fodder which contains zinc additive as an antibiotic. Inhibitory and toxic concentrations are shown in Table 5.

Other organic substances, such as disinfectants, herbicides, pesticides, surfactants, and antibiotics can often flow with the substrate and also cause nonspecific inhibition. All of these substances have a specific chemical formula and it is hard to determine what the behaviour of inhibition will be. Therefore, when such substances do occur in the treated substrate, specific research is strongly advised to determine the concentration of inhibition and possible ways of microorganisms adaptation.

Metal	Inhibition start[1] [mgL-1]	Toxicity to adopted microorganisms[3] [mgL-1]
Cr^{3+}	130	260
Cr^{6+}	110	420
Cu	40	170
Ni	10	30
Cd	70	600
Pb	340	340
Zn	400	600

[1]As inhibitory concentration it is considered the first value that shows diminished biogas production and as toxic concentration it is considered the concentration where biogas production is diminished by 70 %.

Table 5. Inhibitory and toxic concentrations of heavy metals

3. Anaerobic digestion technologies

Block scheme of anaerobic digestion (Fig. 3) shows that technological process of typical anaerobic digestion. It consists of three basic phases: i) substrate preparation and pre-treatment, ii) anaerobic digestion and iii) post treatment of digested material, including biogas use. In this section all of the processes will be elaborated in detail.

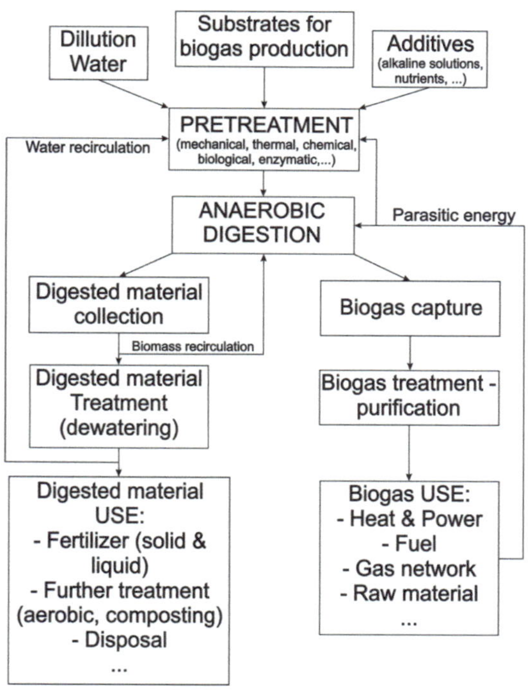

Fig. 3. Block scheme of anaerobic digestion and biogas/digestate utilisation

3.1 Pretreatment

In general, all types of biomass can be used as substrates as long as they contain carbohydrates, proteins, fats, cellulose and hemicellulose as main components. It is however important to consider several points prior to considering the process and biomass pre-treatment. The contents and concentration of substrate should match the selected digestion process. For anaerobic treatment of liquid organic waste the most appropriate concentration is between 2 - 8 % of dry solids by mass. In such case conventional single stage digestion or two stage digestion is used. If considering the treatment of solid waste using solid digestion process, the concentration substrate is between 10 and 20 % by mass. Organic wastes can also contain impurities which usually impairs the process of digestion. Such materials are:

- Soil, sand, stones, glass and other mineral materials
- Wood, bark, card, cork and straw
- Skin and tail hair, bristles and feathers
- Cords, wires, nuts, nails, batteries, plastics, textiles etc.

The presence of impurities in the substrate can lead to increased complexity in the operating expenditure of the process. During the process of digestion of liquid manure from cattle the formation of scum layer on the top of the digester liquid can be formed, caused by straw and muck. The addition of rumen content and cut grass (larger particles than silage) can contribute to its formation. If the substrate consists of undigested parts of corn and grain combined with sand and lime the solid aggregates can be formed at the bottom of the digester and can cause severe clogging problems.

In all such cases the most likely solution is pre-treatment to reduce solids size. Naturally, that all the non-digestible solids (soil, stones, plastics, metals...) should be separated from the substrate flow in the first step. On the other hand grass, straw and fodder residue can contribute to the biogas yield, when properly pretreated, so they are accessible to the digestion microorganisms. Pretreatment can be made by physical, chemical or combined means.

Physical pretreatment is the most common. The best known disintegration methods are grinding and mincing. In grinding and mincing the energy required for operation is inversely proportional to the particle size. Since such energy contributes to the parasitic energy, it should be kept in the limits of positive margin (the biogas yield increased by pre-treatment is more than energy required for it). In the case of organic waste the empirical value for such particle size is between 1 and 4 mm.

Chemical pre-treatment can be used when treating ligno-cellulosic material, such as spent grains or even silage. Very often chemical treatment is used combined with heat, pressure or both. It is common to use acid (hydrochloric, sulphuric or others) or an alkaline solution of sodium hydroxide (in some cases soda or potassium hydroxide). Such solution is added to the substrate in quantities that surpass the titration equilibrium point and then it is heated to the desired temperature and possibly pressurized. Retention times are generally short (up to several hours) compared to retention times of the anaerobic digesters. The pretreated substrate is then much more degradable. The shortage of this pretreatment is low energy efficiency and the cost of chemicals required. It rarely outweighs the costs of building a bigger digester. Therefore it is used mostly in treating industrial waste (such as brewery) where there is plenty of waste lye or acid present and waste heat can be regenerated from the industrial processes as well. Fig. 4 presents the results of our research done on spent brewery grain, where up to 70% of organic matter could be, by means of proper pretreatment, extracted from solid to liquid form, ready for flow-through anaerobic digestion. The research revealed that higher temperatures of pretreatment (120-160°C) enabled finishing of the pretreatment process in 1-2 hours; however the need for a pressurised vessel in such case did not outweigh the time saving.

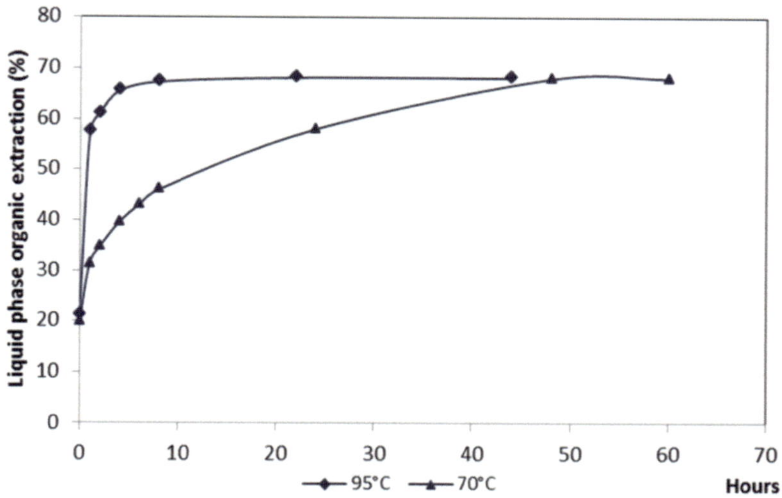

Fig. 4. Effectiveness of thermo-chemical pretreatment

Thermal pretreatment rewards with up to 30 % more biogas production if properly applied. This process occurs at temperature range of 135-220°C and pressures above 10 bar. Retention times are short (up to several hours) and hygienisation is automatically included. Pathogenic microorganisms are completely destroyed. The process runs economically only with heat regeneration. When heat is regenerated from outflow to inflow of the pre-treatment process, it takes only slightly more heat than conventional anaerobic digestion. Such process is very appropriate for cellular material such as raw sewage sludge.

It is also possible to use biological processes as pretreatment. They are emerging in the world. Disintegration takes place by means of lactic acid which decomposes complex components of certain substrates. Recently also disintegration with enzymes has been quite successful, especially using cellulose, protease or carbohydrases at a pH of 4.5 to 6.5 and a retention time of at least 12 days, preferably more (Hendriks and Zeeman 2009).

3.2 Anaerobic digestion

For anaerobic digestion several different types of anaerobic processes and several different types of digesters are applicable. It is hard to say in advance, which digester type is most appropriate for treating the selected organic waste. Digestion of farm waste, for example, should be carried out in decentralized plants to serve each farm separately, to make it an economic and technological unit combined with the farm. In the same sense a town may be a unit in treatment of organic municipal waste. It is important to study the waste of each such unit carefully to be able to determine optimal conditions for substrate digestion. Organic waste can differ very much even in same geographical areas, therefore it is strongly recommended to conduct laboratory and pilot scale experiments before design of the full scale digester is made. Considering the costs of the full scale digester, conducting pilot scale experiments is a minor item, especially if you have no preceding results or experience. The biggest economic setback is when a digester is constructed and it does not perform as expected and consequently requires reconstruction.

There are several processes available to conduct anaerobic digestion. Roughly, the digestion process can be divided into solid digestion and wet digestion processes. Solid digestion processes are in fact anaerobic composters. In this process substrate and biomass are in pre-soaked solid form, containing. 20 % of dry matter and 80 % water. Such processes have several advantages. The main advantage is reducing the reactor volume due to much less water in the system. Four times more concentrated substrate equals approximately four times less reactor volume. It is also possible that some inhibitors (such as ammonium) can have less inhibitory effects in solid digestion process. The biggest disadvantage of solid digestion process is the substrate transport. Substrate in solid form requires more energy for transport in and out of the digesters. It is also a stronger possibility of air intrusion into the digesters, which poses a great risk to process stability and safety. It has been only recently that such processes have gained ground for a wider use. A fine example is the Kompogas® process (Kompogas 2011).

A much larger variety represents wet digestion processes. They operate at conventional concentration up to 5 % of dry solids by mass of the digester suspension. There are several reactor technologies available to successfully conduct anaerobic digestion. Roughly, they can be divided into batch wise (Fig. 5 and Fig. 6) and continuous processes. Furthermore continuous processes can be divided into single stage (Fig. 7) or two-stage processes (Fig. 8). In most of the wet digestion processes microorganisms are completely mixed and suspended with substrate in the digester. The suspended solids of substrate and microorganisms are impossible to separate after the process. If the substrate contains little solids and is mostly dissolved organics liquid, we can apply flow-through processes. In these processes microorganisms are in granules and granules are suspended in liquid which contains dissolved organic material. In such anaerobic processes microorganisms granules are easily separated from the exhausted substrate. Typical representative of such process is the UASB (Upflow Anaerobic Sludge Blanket) process (Fig. 9).

3.2.1 Batch processes

In the batch process all four steps of digestion as well as four stages of treatment process happen in one tank. Typically the reaction cycle of the anaerobic sequencing batch reactor (ASBR) is divided into four phases: load, digestion, settling and unload (Fig. 5). A stirred reactor is filled with fresh substrate at once and left to degrade anaerobically without any interference until the end of the cycle phase. This leads to temporal variation in microbial community and biogas production. Therefore, batch processes require more precise measurement and monitoring equipment to function optimally. Usually these reactors are built at least in pairs, sometimes even in batteries. This achieves more steady flow of biogas for instant use. Between the cycles the tank is usually emptied incompletely (to a certain exchange volume), which is up to 50% of total reactor volume. The residue in the tank serves as microbial inoculum for the next cycle. This makes batch reactors volume larger than of the conventional continuous reactors; however they do not require equalization tanks and the total reactor volume is usually less than in conventional processes. They can be coupled directly to the waste discharge; however this limits the use to more industrial processes (for example food industry) and less to other waste production. Typical cycle time is one day.

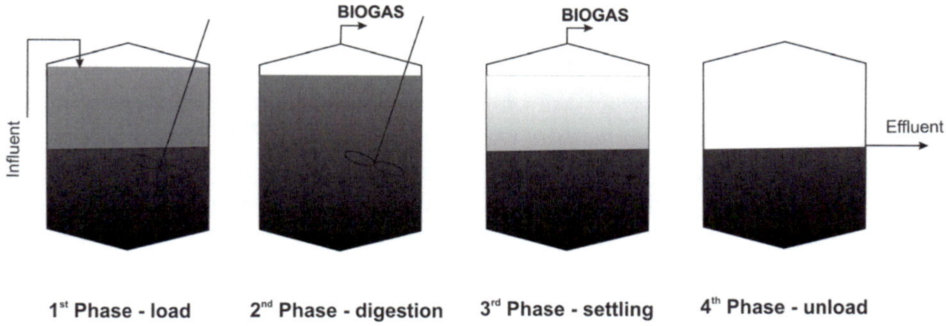

Fig. 5. Schematic picture of the batch ASBR process

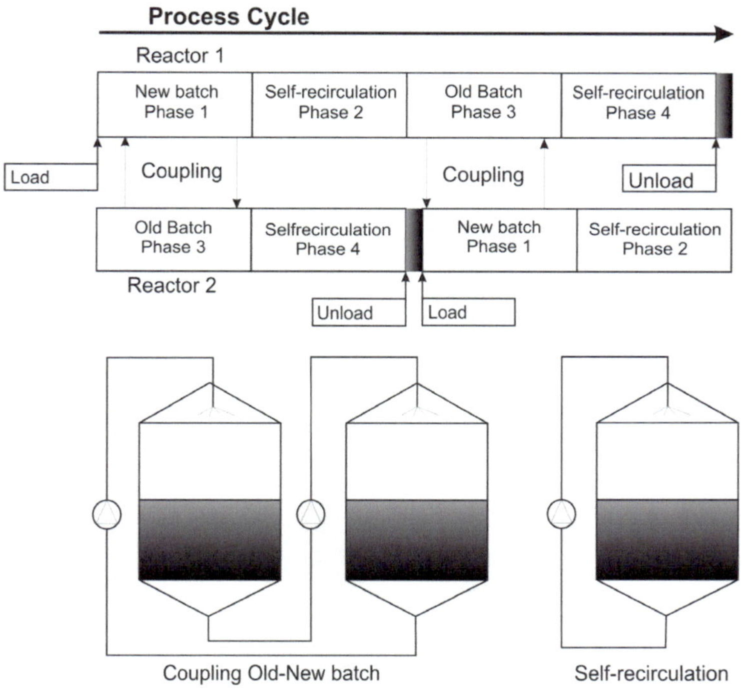

Fig. 6. Batch solid anaerobic digestion

Alternative processes that treat wet organic waste in solid state is reported in literature as SEBAR - Sequential Batch Anaerobic Digester System (Tubtong et al., 2010). In this case the cycle is also divided into four phases, however somehow different than in an ASBR process. This process requires digesters always to be in pairs. The reactor is almost completely emptied between cycles therefore it requires inoculation through leachate exchange between the two digesters (from the one in the peak biogas production to the one at the start of the process). In the other phases leachate is self-circulated (Fig. 6). Typical cycle time is between 30 and 60 days. Although solid substrate reduces the reactor volume, the volume is still rather large due to long cycle times compared to conventional digesters that process liquid substrates. The advantage of this type of digesters is less complicated monitoring equipment so they are applicable in smaller scale.

3.2.2 Continuous processes

Most of the commercial biogas plants use conventional continuous anaerobic digestion process. By conventional it is meant fully mixed, semi-continuous or continuous load and unload reactor at mesophilic temperature range (35-40°C) - Fig. 7. In majority of the cases, the substrate is loaded to the reactor once to several times a day, rarely is it loaded continuously. Continuous load can lead to short circuit, which means that fresh load can directly flow out of the reactor if mixing is too intense or input and output tubes are located improperly. The digester is usually single stage, although they are built in pairs, they do not function as a stage separated process. Usually digesters are equipped with a preparation tank, where various substrates are mixed and prepared for the loading, which also serves as a buffer tank. In many cases also a post-treatment tank is added (it is also called post-fermenter), where treated substrate is completely stabilized and prepared for further treatment. The post-treatment tank can also serve as a buffer towards further treatment steps of the substrate. Generally post-fermenters do not contribute much to overall biogas yield (up to 5 %) if the digester operates optimally. The size of the preparation and post-treatment tanks are determined according to the necessary buffer capacity for continuous operation. The size of the digester is determined with the necessary Hydraulic Retention Time (HRT) and with Organic Loading Rate (OLR) that should be determined in pilot tests. HRT is defined as digester volume divided by substrate flow rate and represents time (in days) in which a certain unit volume of the substrate passes through the reactor. For mesophilic digesters the usual values are between 20 - 40 days, depending on the substrate bio-degradability. In thermophilic digesters to achieve the same treatment efficiency HRT is smaller (between 10 and 20 days).

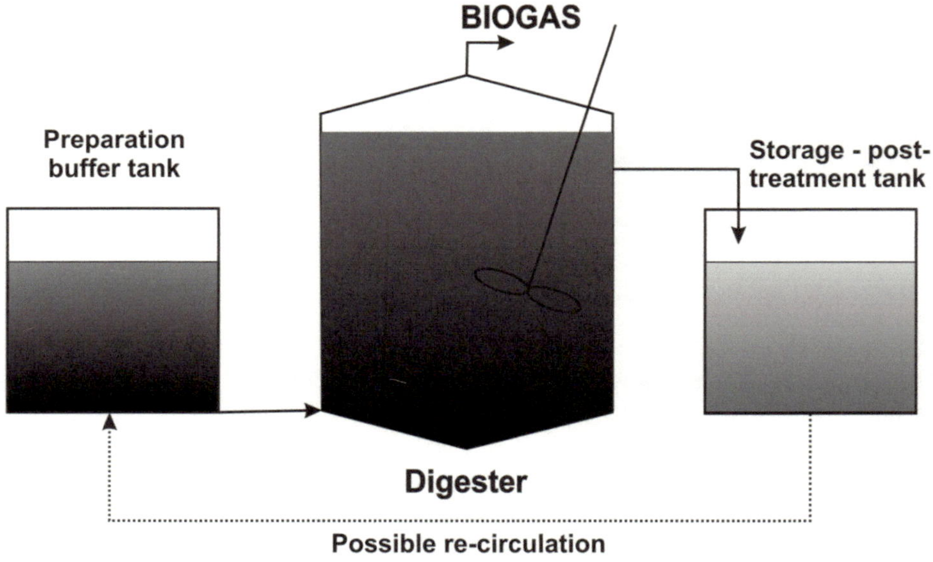

Fig. 7. Conventional single stage anaerobic digestion process

Organic load rate OLR (sometimes also called volume load) is defined as mass of organic material fed to the digester per unit volume per day. Typical value for mesophilic digesters is 2.0-3.0 $kgm^{-3}d^{-1}$. Typical value for thermophilic digesters is 5.0 $kgm^{-3}d^{-1}$. Maximum OLR depends very much of the substrate biodegradability; mesophilic process can rarely achieve higher loads than 5.0 $kgm^{-3}d^{-1}$ and thermophilic 8.0 $kgm^{-3}d^{-1}$. Locally in the digester for a short period of time higher loads can be achieved, however due to inherent instability it is not advisable to run continuously on such high loads.

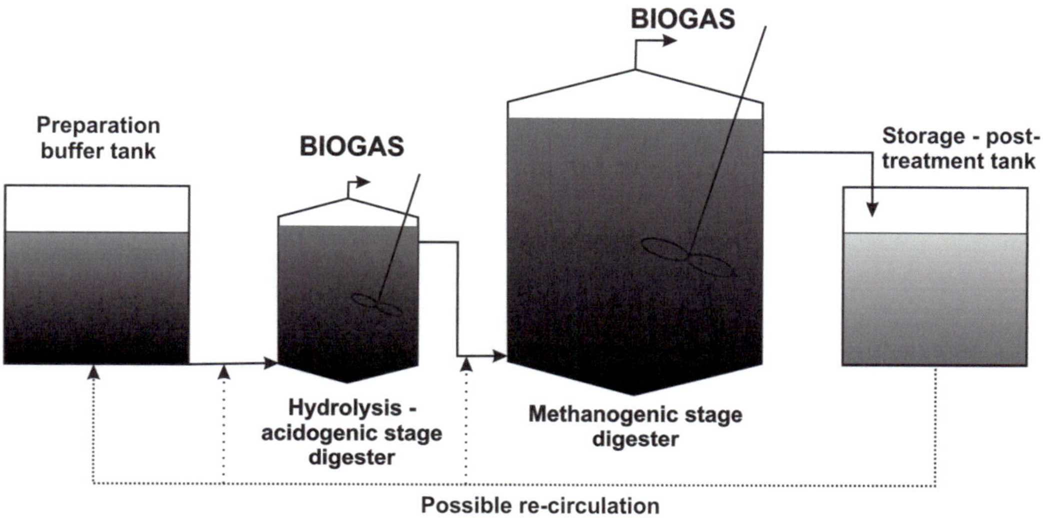

Fig. 8. Two stage anaerobic digestion

To achieve better biodegradation efficiency and higher loads stage separated process can be applied (Fig. 8). In this case the whole substrate or just portions of the substrate which are not easily degradable are treated first in hydrolysis-acidogenic stage reactor and after that in the methanogenic reactor. By separating the biological processes in two separate tanks each can be optimised to achieve higher efficiency with respect to one tank, where all stages of the digestion processes occur simultaneously. Many research data have been published giving considerable attention to this kind of processes (Dinsdale et al., 2000; Song et al., 2004; De Gioannis et al., 2008; Ponsá et al., 2008). Both stages can be either mesophilic or thermophilic, however it is preferred that the hydrolysis-acidogenic reactor is thermophilic and methanogenic is mesophilic. Typical HRT for the thermophilic hydrolysis-acidogenic reactor is 1-4 days, depending on the substrate biodegradability. Typical HRT for the methanogenic reactor is 10 - 15 days (mesophilic) and 10 - 12 days (thermophilic). Advantages of this process beside shorter HRTs are higher organic load rate (20 % or more). Many authors also report slightly better biogas yields (Messenger et al., 1993; Han et al., 1997; Roberts et al., 1999; Tapana and Krishna, 2004). The only disadvantage is more sophisticated equipment and process control, yielding the operation more expensive.

3.2.3 Flow-through processes

The flow-through processes, such as UASB process (Fig. 9) are used only for substrates where most of the organic material is in dissolved form with solids content at maximum 1-5 gL^{-1}. In this substrate category are highly loaded wastewaters of industrial origin (e.g. from beverage industry).

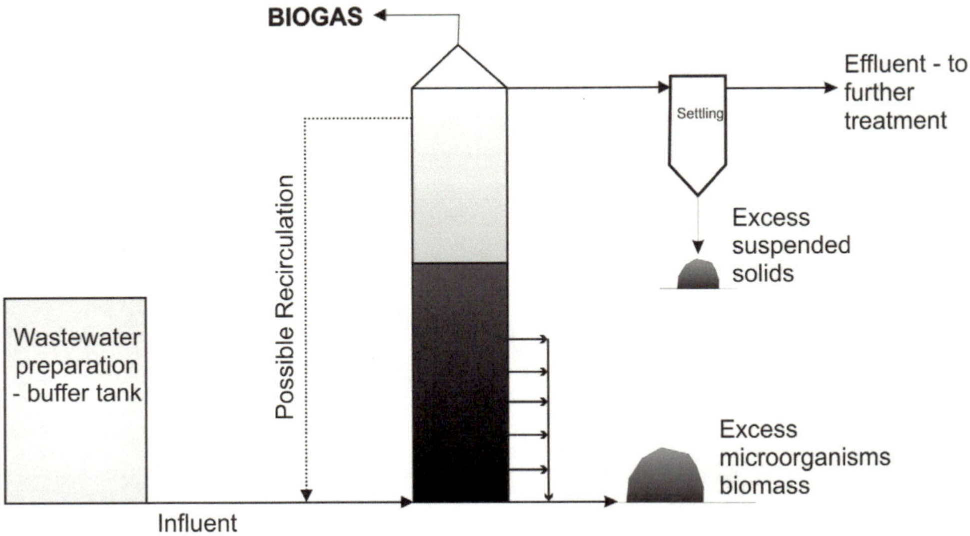

Fig. 9. The UASB process

3.3 Post-treatment and substrate use

After the substrate has been digested, it usually needs additional treatment. There are several possibilities of digested substrate utilisation. Most often, especially in the case of farm waste treatment, the digested substrate is used as a fertilizer. It can be used in liquid state or dewatered. Liquid substrate (total solids concentration 1-5 % by mass) is pumped from the post-treatment tank and spread on the fields. However it must be considered that fertilizing is possible only in certain periods of the year (once or twice). The post-treatment container must be designed accordingly. A possible solution is a lagoon, where digested substrate is stored and additionally stabilized and mineralized during the storing time. When using solid substrate (total solids concentration 20-30 % by mass), the digested substrate is mechanically dewatered first (by belt press or centrifuge) and then liquid and solid parts are used separately. Solid digestate after dewatering can be used fresh as a fertilizer, or it should be stabilized by composting (see further section).

Liquid part of the separated digestate can be used in the new substrate preparation as dilution water, however great caution must be given to nutrients or salts build-up and consequently possible inhibition in the anaerobic digestion. Usually only a portion of that liquid is used in the substrate preparation; the rest must be further treated as a wastewater. Typical concentration of the liquid part of digestate is 200-1000 mgL^{-1} of COD.

3.4 Biogas production, storage, treatment and use

When operating a biogas plant, biogas is the main product and considerable attention must be given to its production, storage, treatment and use. Biogas production completely depends on the efficiency of the anaerobic digestion and its microorganisms. Previous sections have shown what conditions must be met to successfully operate anaerobic digestion. There are two distinct parameters that describe the biogas production:

1. Specific Biogas Productivity - SBP (it's also called biogas yield). It is defined as volume of biogas produced per mass of substrate inserted into digester (m^3kg^{-1}). There are variations; SBP can be expressed in m^3 of gas per kg of substrate: i) (wet) mass, ii) total solids, iii) volatile organic solids or iv) COD. SBP tells us how much biogas was produced from the chosen unit of substrate. Maximum possible SBP for certain substrate is called biogas potential. Biogas potential can be determined by a standard method (ISO 1998).
2. Biogas Production Rate – BPR. It is defined as volume of biogas produced per volume of the digester per day ($m^3m^{-3}d^{-1}$). BPR tells us how much biogas we can gain from the active volume of a digester in one day.

SBP values of an optimally operating digester reach 80-90 % of the biogas potential. Typical values of SBP for farm waste are shown in Table 1. Typical values of BPR for mesophilic digesters are from 0.9 to 1.3 $m^3m^{-3}d^{-1}$. Lower values indicate the digester is oversized; higher values are rare or impossible, due to anaerobic process failure. For thermophilic or two stage digesters the typical BPR values are from 1.3 to 2.1 $m^3m^{-3}d^{-1}$, respectively. UASB reactors are much less volume demanding and can achieve a BPR of up to 10 $m^3m^{-3}d^{-1}$.

Biogas production is rarely constant; it is prone to fluctuations due to variation of loading rates, inner and outer operating conditions, possible inhibitions etc... Therefore, a buffer volume is required for the biogas storage. This enables the biogas user to get a constant biogas flow and composition. Most of the modern biogas plants are equipped with co-generation units (named also combined heat and power units - CHP) which require constant gas flow for steady and efficient operation. There are several possibilities of biogas storage; roughly they can be divided into low pressure (10-50 mbar) and high pressure storage (over 5 bar). Low-pressure storage is used in on-site installations and for gas grid delivery; high pressure storage is used for long term storage, for transport in high pressure tanks and in installations with scarce space for volume extensive low pressure holders.

Low pressure biogas holders arise in many variations. It is possible to include biogas holder in the design of the digester. The most known is the digester with a movable cover. These digesters are less common, because a movable cover requires increased investment and operating expenditure. More common are external biogas holders that are widely commercially available. An example of a modern biogas holder is presented in Fig. 10.

Low pressure biogas holders require an extensive volume of 30 to 2000 m^3 (Deublin and Steinhauser, 2008). Usually the pressure is kept constant and the volume of the bag is varied. High pressure biogas holders are of constant volume and made of steel, they are subject to special safety requirements. They do require more complex equipment for compression and expansion of the gas and are more cost-effective for operation and maintenance.

Biogas contains methane (40-70% by volume) and carbon dioxide. There are also components, present in low concentrations (below 1 %) such as water vapour, substrate micro particles and trace gases. Therefore biogas treatment is necessary to preserve equipment for its storage, transport and utilisation. Solid particles can be filtered out by candle filters, sludge and foam is separated in cyclones. For removal of trace gases, where hydrogen sulphide (H_2S) is the most disturbing one due to its corrosion properties, processes like scrubbing, adsorption and absorption are used. In some cases also drying is required (usually to the relative humidity of less than 80 %).

After cleaning, biogas is used to produce energy. The most common way is to us all biogas in cogeneration plant in CHP unit to produce power and heat simultaneously (Fig. 11). In this case we can achieve maximum power production and enough excess heat to run the digesters. The energy required for heating the digester is also called parasitic energy. The anaerobic digesters require heat to bring the substrate to operating temperature and to compensate the digester heat losses. The digester also requires energy for mixing, substrate pumping and pre-treatment. The largest portion of heating demands in the digester operation is substrate heating. It requires over 90 % of all heating demands, and only up to 10 % is required for heat loss compensation (Zupancic and Ros 2003). In mesophilic digestion a CHP unit delivers enough heat for operation, while in thermophilic digestion additional heat is required. This additional heat demand can be covered by heat exchange between substrate outflow to substrate inflow -

Usually a conventional counter-currant heat exchanger is sufficient; however a heat pump can be applied as well.

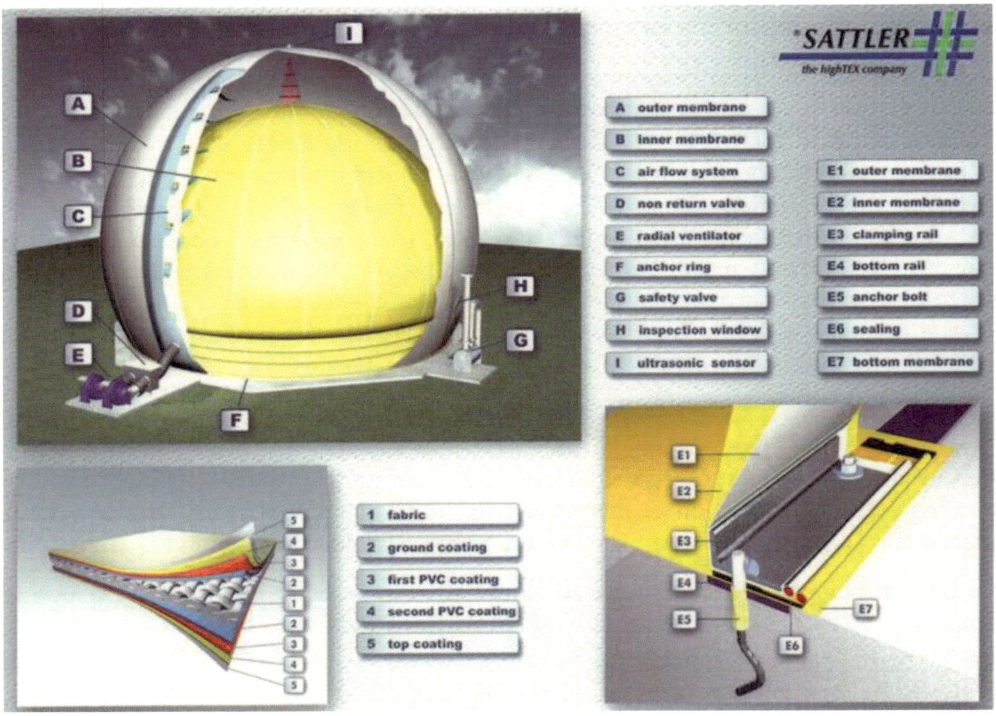

Fig. 10. An example of commercially available biogas holder (Sattler 2011)

Electric energy is also required in digester operation for pumping, mixing and process control and regulation. In practice, no more than 10-15 % of electric energy produced should be used for internal demands.

The pre-treatment process may also require electric or thermal energy. Pre-treatment improves anaerobic digestion and its biogas production. However implications of pre-treatment methods must be carefully considered. The golden rule is that pre-treatment should not spend more energy that it helps to produce. If the energy use and production balances out, pre-treatment may have benefits such as more stable digested substrate, smaller digesters, pathogen removal etc. There are substrates that require extensive pre-treatment; especially this is the case for ligno-cellulosic material (like spent brewery grains, Sežun et al. 2011) that require energy intensive pre-treatment to be successfully digested. In such cases the energy need for pre-treatment must be accounted in the energy production. In many cases it cannot outweigh the economy of the process; it may well happen that the parasitic energy demand is too high.

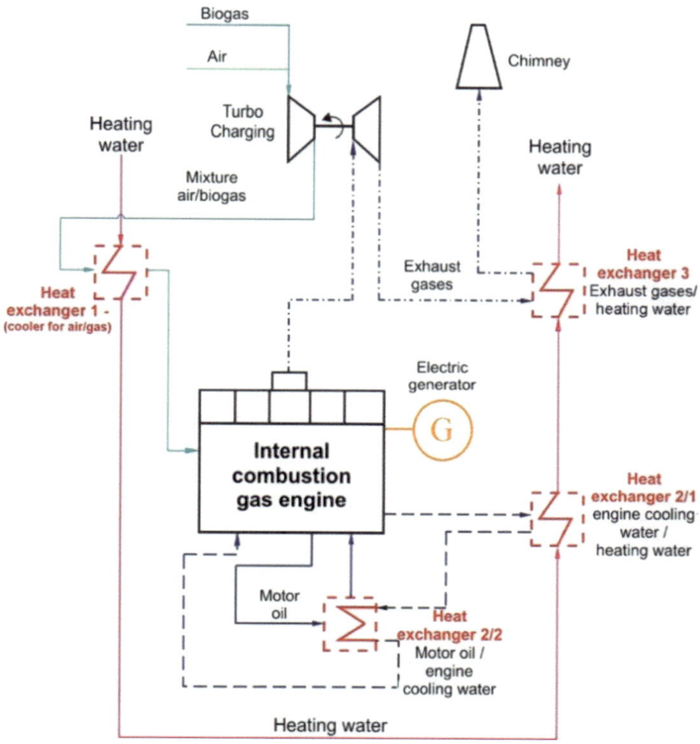

Fig. 11. Schematic of a combined heat and power units (CHP) unit

In recent years great interest was taken to biogas injection into natural gas grid. Mainly due to the fact, that global energy efficiency in such cases is usually far greater that at CHP plants. Namely in warmer periods of the year, heat produced in the CHP is largely wasted and therefore unused. Injecting the biogas into natural gas grid assures more than 90% energy efficiency, due to the nature of the use (heat production), even in warmer periods. Consequently the whole biogas production process can be more economic, in some cases even without considerable subsidies as well as more renewable energy is put to the energy supply. Also in most cases, the investment costs of biogas plants may be less, since there is no CHP plant. In order to be able to inject the biogas into natural gas as biomethane (Ryckebosch et al., 2011) grid certain purity standards must be fulfilled, which in EU are determined by national ordinances (a good example is the German ordinance for Biogas injection to natural gas grids from 2008), where responsibilities of grid operators and biogas producers are determined (Fig. 13) as well as quality standards are prescribed (DVGW, 2010). When injecting biomethane into the natural gas grid some biogas must be used for the reactors self-heating. It is advisable to use regeneration (Fig. 12), even in mesophilic temperature ranges, to minimize this expenditure, which on annual basis can contribute to 10-20 % of all biogas production (Pöschl et al., 2010).

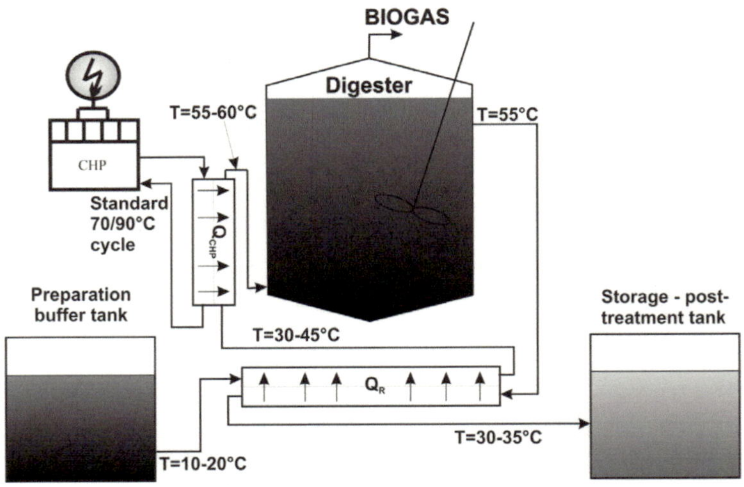

Fig. 12. Scheme of the heat regeneration from output to input flows.

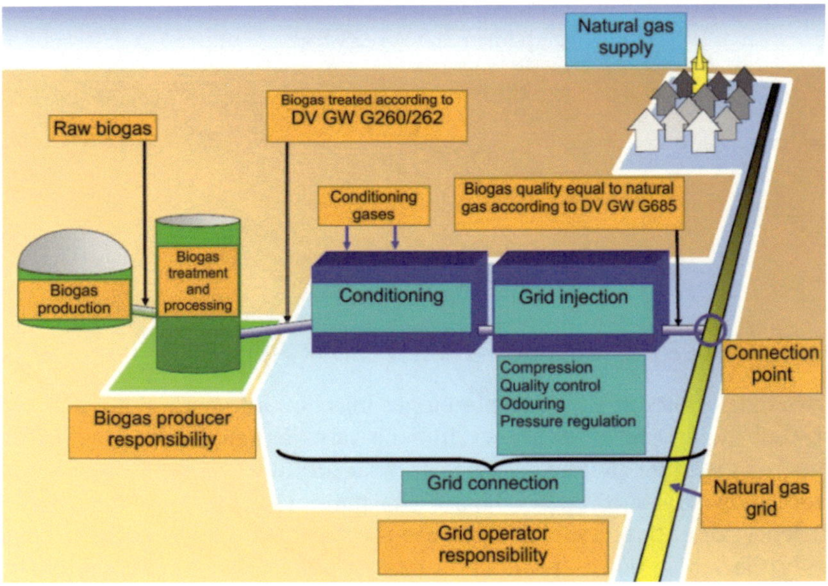

Fig. 13. Injection of biogas into natural gas grid (Behrendt and Sieverding 2010)

3.5 Anaerobic digestion residue management

Environmental impact assessment of an anaerobic digestion plant (a biogas station) should take into consideration both the plant emissions and the digestate management. The first aspect mainly relates to flue gas and odour emissions. The exhaust gases from gas motors must fulfil emission limit values, which is not a problem when appropriate gas pretreatment (sulphide and ammonia removal) has been applied. Unpleasant odours mainly originate from storage, disintegration and internal transport of organic waste. These should be carried over in a closed system, equipped with an air collection system fitted with a biofilter or connected with the gas motor air supply.

3.5.1 The anaerobic digestion residue management

A quality management system (QMS) specific to a defined digestion process and its resulting whole digestate or any separated liquor and separated fibre, should be established and maintained. Anaerobically digested slurry or sludge contains 2-12 % of solids; wet waste from solid state digestion contains 20-25 % solids. The digestate contains not degraded organic waste, microorganism cells and structures formed during digestion, as well as some inorganic matter. This is potentially an alternative source of humic material, nutrients and minerals to the agricultural soil (PAS, 2010). It may be used directly or separated into liquid and solid part. The liquid digestate is often recycled to the digestion process; some pretreatment may be required to reduce nitrogen or salt content.. Freshly digested organic waste is not stable under environmental conditions: it has an unpleasant odour, contains various noxious or corrosive gases such as NH_3 and H_2S, and still retains some biodegradability. In certain periods of a year it may be used in agriculture directly, in most cases however it must be stabilized before being applied to the fields.

Aerobic treatment (composting) is an obvious and straightforward solution to this problem. The composting procedure has several positive effects: stabilization of organic matter, elimination of unpleasant odours and reduction of pathogenic microorganisms to an acceptable level. Composting, applied prior to land application of the digested waste, contributes also to a beneficial effect of compost nitrogen availability in soil. (Zbytniewski and Buszewski, 2005; Tarrasón et al., 2008)

The simplest way is composting of the dehydrated fresh digestate in a static or temporarily turned-over pile. A structural material is necessary to provide sufficient porosity and adequate air permeability of the material in the pile. Various wood or plant processing residues may be used as a structural material like woodchips, sawdust, tree bark, straw and corn stalks provided that the sludge : bulk agent volume ratio is between 1:1 and 1:4 (Banegas et al., 2007). The majority of organic material is contributed by the bulking agent, but significant biodegradation of the digestate organic material also occurs, by means of natural aerobic microorganisms.

The final compost quality depends on the content of pollutants such as heavy metals, pathogenic bacteria, nutrients, inert matter, stability etc. in the mature compost. Typical quality parameters are presented in Table 6. The properties of the compost standard leachate may also be considered. Heavy metals and persistent organic pollutants accumulate in the compost and may cause problems during utilization. Compost quality depends on quality of the input material, which should be carefully controlled by input analysis. Pathogenic bacteria may originate from the mesophilic digestates or from infected co-composting materials, if applied (e.g. food waste). If thermophilic phase period of the composting process has lasted at least few days, the compost produced may be considered sanitized and free of pathogens such as *Salmonella, Streptococci* and *coliforms.*

The third important factor is presence of nitrogen. Several authors have reported that the optimal C/N ratio is between 25/1 and 30/1 although operation at low C/N ratios of 10/1 are also possible. With such low C/N ratios the undesirable emission of ammonia can be significant (Matsumura et al., 2010). Characteristic values of organic matter content and total nitrogen in the digested sludge are 50-70% and 1.5-2.5%, respectively. In the first week of the digested sludge composting the total carbon is reduced by between 11% and 27% and total nitrogen is reduced by between 13% and 23% (Pakou et al., 2009; Yañez et al., 2009).

Parameter	Test method	Limit value
1. Pathogens		
Escherichia coli	ISO 16649-2	1000 CFU/g fresh matter
Salmonella sp.		Absent in 25 g fresh matter
2. Toxic elements		
Cd	EN 13650	1.5 mg/kg d.m.
Cr	EN 13650	100 mg/kg d.m.
Cu	EN 13650	200 mg/kg d.m.
Hg	ISO 16772	1.0 mg/kg d.m.
Ni	EN 13650	50 mg/kg d.m.
Pb	EN 13650	200 mg/kg d.m.
Zn	EN 13650	400 mg/kg d.m.
3. Stability		
Volatile faty acids	GC	-
Residual biogas potential		0.25 l/gVS
4. Physical contaminants		
Total glass, metal, plastic and other man-made fragments		0.5 $\%_{m/m}$ d.m.
Stones >5 mm		8 $\%_{m/m}$ d.m.
5.Parameters for declaration		
Total nitrogen	EN 13654	
Total phosphorus	EN 13650	
Total potassium	EN 13650	
Water soluble chloride	EN 13652	-
Water soluble sodium	EN 13652	
Dry matter	EN 14346	
Loss on ignition	EN 15169	
pH	EN 13037	

Table 6. Control parameters of digestate quality for application in agriculture (WRAP 2010)

Highest degradation rates in the compost pile are achieved with air oxygen concentration above 15% which also prevents formation of anaerobic zones. The quality of aeration depends primarily on structure and degree of granulation of the composting material; finer materials generally provide better aeration of the compost pile (Sundberg and Jönsson, 2008). In the first stages of degradation, acids are generated, and these tend to decrease the pH in the compost pile. The optimum pH range for microorganisms to function is between 5.5 and 8.5. Elevated temperature in the compost material during operation is a consequence of exothermic organic matter degradation process. The optimum temperature for composting operation, in which pathogenic microorganisms are sanitised, is 55-70°C. In the initial phases of composting the prevailing microorganisms are fungi and mesophilic bacteria, which contribute to the temperature increase and are mostly sanitised in the relevant thermophilic range. When temperature falls many of the initial mesophilic microorganisms reappear, but the predominant population are more highly evolved organisms such as protozoa and arthropods (Schuchard, 2005). For optimum composting operation the correct conditions must be established and are determined by particle size distribution and compost pile aeration have shown that the air gaps in the compost pile can be reduced from an initial 76.3% to a final 40.0%. The optimum moisture content in the compost material is in the range of 50-70%.

In the recent years the composting practice for anaerobic digestate has been thoroughly studied for many different types of substrates, for co-composting and with many different bulk agents (Nakasaki et al., 2009; Himanen et al., 2011).

From various reasons the composting of the digestate residue is sometimes not possible (lack of space, problems with compost disposal etc.). Alternatively the digestate may be treated by thermal methods, which require higher solid content. Mechanical dehydration by means of continuous centrifuges provides solid content about 30 % with positive calorific value. Incineration may be carried out in a special kiln (most often of fluidized bed type) or together with municipal waste in a grit furnace. Co-incineration in industrial kilns usually require drying of sludge to 90 % dryness, that gives calorific value of about 10 MJ/kg. Thermal methods are more expensive than composting due to high energy demand for dehydration and drying, sophisticated processes invplved and strict monitoring requirements. Good review of the modern alternative processes of anaerobic sludge treatment is presented by Rulkens (2008).

4. Conclusions

The chapter entitled "Sustainable Treatment of Organic Wastes" presents principles and techniques for treatment of wet biodegradable organic waste, which can be applied in order to achieve environmental as well as economic sustainability of their utilisation.

The chapter mostly focuses on organic wastes generated in the municipal sector; however it may well apply to similar wastes from agriculture and industry. The main focus is aimed at matching the anaerobic treatment process to the selected type of waste in order to maximize the biogas production, a valuable renewable energy resource. The chapter also focuses on technological aspects of the technology used in such treatments and presents and elaborates several conventional treatments (such as semi-continuous processes, two stage processes, sequencing batch processes, etc.) as well as some emerging technologies which have only recently gained some ground (such as anaerobic treatment in solid state). The basic conditions are presented which are required to successfully design and operate the treatment process. Organic loading rates, biogas production rates, specific biogas productivity, biogas potentials and specific concerns for certain technologies and waste substrates are presented. The main influencing factors such as environmental conditions (pH, temperature, alkalinity, etc.) have been addressed as well as inhibitors that can arise in such processes (heavy metals, ammonia, salts, phenolic compounds from lignocellulosic degradation, organic overload etc.). The biogas treatment and use, such as power production and natural gas grid injection, have been presented as well as the use of parasitic energy, options for biogas production enhancement through waste pre-treatment (mechanical, chemical, physical, etc.) and treatment of residues of anaerobic digestion, which may have an important impact on the environment. Special attention is given to further treatment of digested solid residues as well. Due attention is paid to aerobic stabilization processes (open and closed composting), taking into account physical form of the waste, its composition, pollution, degradability and final deposition and use.

5. Acknowledgment

The authors express acknowledgements to Slovenian biogas producers and to the Slovenian Science and Research Agency, whose support in anaerobic digestion and waste treatment research has lead to the knowledge presented here.

6. References

Banegas V., J.L. Moreno, J.I. Moreno, C. Garcia, G. Leon, T. Hernandez, (2007). Composting anaerobic and aerobic sewage sludges using two proportions of sawdust. Waste Manag. 27, 1317-27

Behrendt F.; Sieverding M., (2010). "Biogas-Einspeisung in Erdgasnetze aus Netzbetreibersicht", GWF Das Gas- und Wasserfach, Gas - Erdgas, 151(3), p. 140-144.

De Gioannis G., Diaz L.F., Muntoni A., Pisanu A., (2008). "Two-phase anaerobic digestion within a solid waste/wastewater integrated management system." Waste Management 28(10): 1801-1808.

Deublein, D. and A. Steinhauser (2008). Biogas from waste and renewable resources. Weinheim, Willey-VCH Verlag GmbH & Co. KGaA.

Dinsdale R. M., Premier G.C., Hawkes F.R., Hawkes D.L., (2000). "Two-stage anaerobic co-digestion of waste activated sludge and fruit/vegetable waste using inclined tubular digesters." Bioresource Technology 72(2): 159-168.

DVGW - Deutscher Verein des Gas- und Wasserfaches, DVGW G 262 "Nutzung von Gasen aus regenerativen Quellen in der der Öffentlichen Gasversorgung"(2010), available online with limited access at http://www.dvgw.de/gas/gesetze-und-erordnungen/.

Feijoo G., Soto M., Méndez R., Lema J.M., (1995). "Sodium inhibition in the anaerobic digestion process: antagonism and adaptation phenomena." Enzym Microb Technol 17(2): 180-188.

GNS (2009) Nitrogen removal from manure and organic residues by ANAStrip - process (System GNS). http://www.gns-halle.de/english/site_1_6.htm (Access 11th August 2011).

Han Y., Sung S., Dague R.R., (1997). "Temperature-phased anaerobic digestion of wastewater sludges." Water Sci Technol 36(6-7): 367-374.

Hendriks A.T.W.M., Zeeman G., (2009). "Pretreatments to enhance the digestibility of lignocellulosic biomass." Bioresour Technol 100(1): 10-18.

Himanen M., Hänninen K., (2011). Composting of bio-waste, aerobic and anaerobic sludges–Effect of feedstock on the process and quality of compost. Bioresource Technology, Volume 102, Issue 3, p. 2842-2852.

ISO (1998) EN ISO 11734 (1998) Water Quality - Evaluation of the »Ultimate« Anaerobic Biodegradability of Organic Compounds in Digested Sludge - Method by Measurement of the Biogas Production, International Standard Organization.

Kompogas, (2011). " Green energy from organic waste." Retrieved 11th August 2011, from http://www.axpo-kompogas.ch/index.php?path=home&lang=en.

Messenger J., de Villers H.A., Laubscher S.J.A., Kenmuir K. and Ekama G.A. (1993). "Evaluation of the Dual Digestion System: Part 1: Overview of the Milnerton Experience." Water SA 19(3): 185-192.

Mrafkova L., Goi D., Gallo V., Colussi I., (2003). "Preliminary Evaluation of Inhibitory Effects of Some Substances on Aerobic and Anaerobic Treatment Plant Biomasses." Chem Biochem Eng Q 17(3): 243-247.

Nakasaki K., Tran L.T.H., Idemoto Y., Abe M., Rollon A.P., (2009). Comparison of organic matter degradation and microbial community during thermophilic composting of two different types of anaerobic sludge. Bioresource Technology, Volume 100, Issue 2, p. 676-682.

Polprasert C. (2007) Organic Waste Recycling - Technology and Management, 3rd Ed., e-book, IWA Publishing , London http://books.google.si/books?id=owycqJMjoZoC&printsec=frontcover&dq=Organic+waste+recycling:+technology+and+management&source=bl&ots=kZC9lGSJFb&sig=T_rGd1UQVQL3dLDmbacV87jNXEQ&hl=sl&ei=-

Ponsá S., Ferrer I., Vázquez F., Font X. (2008). "Optimization of the hydrolytic-acidogenic anaerobic digestion stage (55 °C) of sewage sludge: Influence of pH and solid content." Water Res 42(14): 3972-3980.

Pöschl M., Ward S., Owende P., (2010). "Evaluation of energy efficiency of various biogas production and utilization pathways"Applied Energy, 87 (11), p. 3305-3321.

Roberts R., Davies W.J., Forster C.F., (1999). "Two-Stage, Thermophilic-Mesophilic Anaerobic Digestion of Sewage Sludge." Process Saf Environ Protec 77(2): 93-97.

Rulkens W., (2008). Sewage Sludge as a Biomass Resource for the Production of Energy: Overview and Assessment of the Various Options, *Energy Fuels, 22* (1), pp 9–15

Ryckebosch E., Drouillon M., Vervaeren H., (2011). "Techniques for transformation of biogas to biomethane", Biomass and Bioenergy, 35(5), p. 1633-1645.

Sattler. (2009). "Biogas storage systems." Retrieved 11.8.2011, from http://www.sattler-ag.com/sattler-web/en/products/190.htm

Schuchard F., (2005). Composting of organic waste. In: Enviromental biotechnology concepts and aplications. Editors:Jördering, H.J., Winter, J. Weinheim: Willey-VCH Verlag, p. 333-354.

Sežun M., Grilc V., Zupančič G.D. and Marinšek-Logar R., (2011). Anaerobic digestion of brewery spent grain in a semi-continuous bioreactor: inhibition by phenolic degradation products. Acta chim. *slov.., 58, 1, 158-166.*

Song Y.C., Kwon S.J., Woo J.H., (2004). "Mesophilic and thermophilic temperature co-phase anaerobic digestion compared with single-stage mesophilic- and thermophilic digestion of sewage sludge." Water Research 38(7): 1653-1662.

Sundberg C., Jönsson H., (2008). Higher pH and faster decomposition in biowaste compostingby increased aeration. Waste Management, Volume 28, Issue 3, p. 518-526.

Sung, S. and T. Liu (2003). "Ammonia inhibition on thermophilic anaerobic digestion." Chemosphere 53: 43-52.

Tubtong C., Towprayoon S., Connor M.A., Chaiprasert P., Nopharatana A., (2010). "Effect of recirculation rate on methane production and SEBAR system performance using active stage digester." Waste Management & Research, 28(9): 818-827.

Tapana C., Krishna P.R., (2004). "Anaerobic Thermophilic/Mesophilic Dual-Stage Sludge Treatment." Journal of Environmental Engineering 126(9): 796-801.

Tarrasón, D., Ojeda, G., Ortiz, O., Alcañiz, J.M., (2008). Differences on nitrogen availability in a soil amended with fresh, composted and thermally-dried sewage sludge. Bioresource Technology, Volume 99, Issue 2, p. 252-259.

Verordnung zut Anderung der Gesetzzugangsverordnung, das Gasnetzentgeltverodnung, der Anreizregulierungsverordnung und der Stromnetzentgeltverordnung, Bundesgesetzblatt Jahrgang 2008 Teil I Nr. 14 11. April 2008 (S. 693÷697)

WRAP (2010) Specification for whole digestate, separated digestate, separated liquor and separated fibre derived from the anaerobic digestion of source-segregated biodegradable materials, Publicly Available Specification (PAS) 110, U.K.

Yañez R., Alonso J.L., Díaz M.J., (2009). Influence of bulking agent on sewage sludge composting process, Bioresource Technology, Volume 100, Issue 23, p. 5827-5833.

Ye C., Cheng J.J., Creamer K.S., (2008). "Inhibition of anaerobic digestion process: A review." Bioresource Technology 99(10): 4044-4064.

Zupančič G. D., Roš M., (2003). "Heat and energy requirements in thermophilic anaerobic sludge digestion." Renewable Energy 28(14): 2255-2267.

Zbytniewski, R., Buszewski, B., 2005. Characterization of natural organic matter (NOM) derived from sewage sludge compost. Part 1: chemical and spectroscopic properties. Bioresource Technology, Volume 96, Issue 4, p. 471-478.

2

Vermicomposting: Composting with Earthworms to Recycle Organic Wastes

Jorge Domínguez[1] and María Gómez-Brandón[2]
[1]Universidade de Vigo
[2]University of Innsbruck
[1]Spain
[2]Austria

1. Introduction

The overproduction of organic wastes has led to the use of inappropriate disposal practices such as their indiscriminate and inappropriately-timed application to agricultural fields. These practices can cause several environmental problems, including an excessive input of potentially harmful trace metals, inorganic salts and pathogens; increased nutrient loss, mainly nitrogen and phosphorus, from soils through leaching, erosion and runoff; and the emission of hydrogen sulphide, ammonia and other toxic gases (Hutchison et al., 2005). However, if handled properly, organic wastes can be used as valuable resources for renewable energy production, as well as sources of nutrients for agriculture, as they provide high contents of macro- and micronutrients for crop growth and represent a low-cost alternative to mineral fertilizers (Moral et al., 2009).

The health and environmental risks associated with the management of such wastes could be significantly reduced by stabilizing them before their disposal or use. Composting and vermicomposting are two of the best known-processes for the biological stabilization of a great variety of organic wastes (Domínguez & Edwards, 2010a). However, more than a century had to pass until vermicomposting, i.e. the processing of organic wastes by earthworms was truly considered as a field of scientific knowledge or even a real technology, despite Darwin (1881) having already highlighted the important role of earthworms in the decomposition of dead plants and the release of nutrients from them.

In recent years, vermicomposting has progressed considerably, primarily due to its low cost and the large amounts of organic wastes that can be processed. Indeed, it has been shown that sewage sludge, paper industry waste, urban residues, food and animal waste, as well as horticultural residues from cultivars may be successfully managed by vermicomposting to produce vermicomposts for different practical applications (reviewed in Domínguez, 2004). Vermicompost, the end product of vermicomposting, is a finely divided peat-like material of high porosity and water holding capacity that contains many nutrients in forms that are readily taken up by plants.

Vermicomposting is defined as a bio-oxidative process in which detritivore earthworms interact intensively with microorganisms and other fauna within the decomposer community, accelerating the stabilization of organic matter and greatly modifying its physical and biochemical properties (Domínguez, 2004). The biochemical decomposition of organic matter is primarily accomplished by microorganisms, but earthworms are crucial drivers of the process as they may affect microbial decomposer activity by grazing directly on microorganisms (Aira et al., 2009; Monroy et al., 2009; Gómez-Brandón et al., 2011a), and by increasing the surface area available for microbial attack after comminution of organic matter (Domínguez et al., 2010) (Figure 1). These activities may enhance the turnover rate and productivity of microbial communities, thereby increasing the rate of decomposition. Earthworms may also affect other fauna directly, mainly through the ingestion of microfaunal groups (protozoa and nematodes) that are present within the organic detritus consumed (Monroy et al., 2008); or indirectly, modifying the availability of resources for these groups (Monroy et al., 2011) (Figure 1).

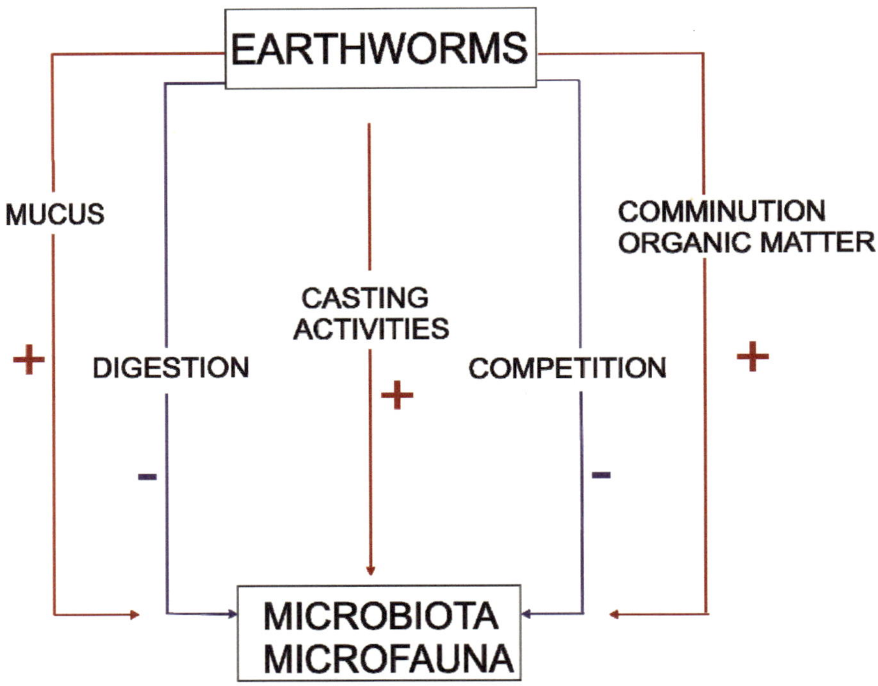

Fig. 1. Positive (+) and negative (-) effects of earthworms on microbiota and microfauna (modified from Domínguez et al., 2010).

Furthermore, earthworms are known to excrete large amounts of casts (Figure 1), which are difficult to separate from the ingested substrate (Domínguez et al., 2010). The contact between worm-worked and unworked material may thus affect the decomposition rates (Aira & Domínguez, 2011), due to the presence of microbial populations in earthworm casts different from those contained in the material prior to ingestion (Gómez-Brandón et al., 2011a). In addition, the nutrient content of the egested materials differs from that in the ingested material (Aira et al., 2008), which may enable better exploitation of resources, because of the presence of a pool of readily assimilable compounds in the earthworm casts. Therefore, the decaying organic matter in vermicomposting systems is a spatially and temporally heterogeneous matrix of organic resources with contrasting qualities that result from the different rates of degradation that occur during decomposition (Moore et al., 2004).

2. Earthworm species suitable for vermicomposting

Earthworms represent the major animal biomass in most terrestrial temperate ecosystems (Edwards & Bohlen, 1996). Indeed, more than 8,300 species of earthworms have been described (Reynolds & Wetzel, 2010), although for the great majority of these species only the names and morphologies are known, and little is yet known about their biology, life cycles and ecology. Different species of earthworms have different life histories, occupy different ecological niches, and have been classified, on the basis of their feeding and burrowing strategies, into three ecological categories: epigeic, anecic and endogeic (Bouché 1977). Endogeic species (soil feeders) forage below the surface soil, ingest high amounts of mineral soil and form horizontal burrows. Anecic species (burrowers) live in deeper zones of mineral soils, ingest moderate amounts of soil, and feed on litter that they drag into their vertical burrows. And, epigeic earthworms (litter dwellers and litter transformers) live in the soil organic horizon, in or near the surface litter, and mainly feed on fresh organic matter contained in forest litter, litter mounds and herbivore dungs, as well as in man-made environments such as manure heaps. These latter species, with their natural ability to colonize organic wastes; high rates of consumption, digestion and assimilation of organic matter; tolerance to a wide range of environmental factors; short life cycles, high reproductive rates, and endurance and resistance to handling show good potential for vermicomposting (Domínguez & Edwards, 2010b). In fact, few epigeic earthworm species display all these characteristics, and only four have been extensively used in vermicomposting facilities: *Eisenia andrei*, *Eisenia fetida*, *Perionyx excavatus* and *Eudrilus eugeniae* (Figure 2).

Fig. 2. Earthworm species *Eisenia andrei* (top left), *Eisenia fetida* (top right), *Eudrilus eugeniae* (bottom left) and *Perionyx excavatus* (bottom right).

3. How does vermicomposting work?

The vermicomposting process includes two different phases regarding earthworm activity: (i) an active phase during which earthworms process the organic substrate, thereby modifying its physical state and microbial composition (Lores et al., 2006), and (ii) a maturation phase marked by the displacement of the earthworms towards fresher layers of undigested substrate, during which the microorganisms take over the decomposition of the earthworm-processed substrate (Aira et al., 2007; Gómez-Brandón et al., 2011b). The length of the maturation phase is not fixed, and depends on the efficiency with which the active phase of the process takes place, which in turn is determined by the species and density of earthworms (Domínguez et al., 2010), and the rate at which the residue is applied (Aira & Domínguez, 2008).

More specifically, the impact of earthworms on the decomposition of organic waste during the vermicomposting process is initially due to *gut associated processes* (GAPs) (Figure 3), i.e., via the effects of ingestion, digestion and assimilation of the organic matter and microorganisms in the gut, and then casting (Gómez-Brandón et al., 2011a). Specific microbial groups respond differently to the gut environment (Schönholzer et al., 1999) and selective effects on the presence and abundance of microorganisms during the passage of organic material through the gut of these earthworm species have been observed. For instance, some bacteria are activated during passage through the gut, whereas others remain

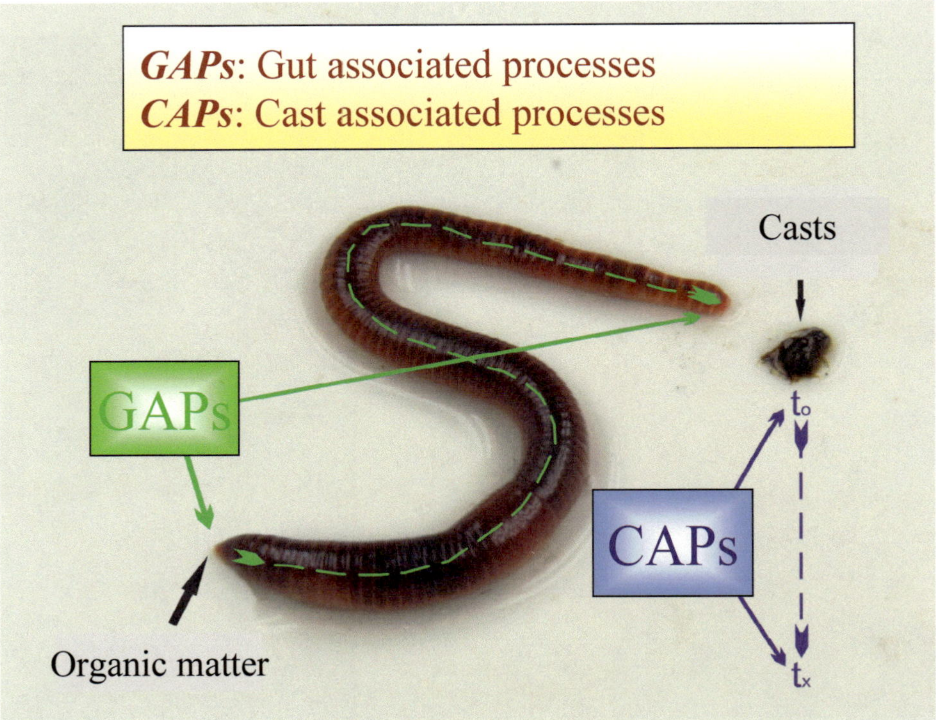

Fig. 3. Earthworms affect the decomposition of organic matter during vermicomposting through ingestion, digestion and assimilation in the gut and then casting (*gut associated processes*); and *cast associated processes,* which are more closely related with ageing processes.

unaffected and others are digested in the intestinal tract and thus decrease in number (Drake & Horn, 2007; Monroy et al., 2009). Such selective effects on microbial communities as a result of gut transit may alter the decomposition pathways during vermicomposting, probably by modifying the composition of the microbial communities involved in decomposition, as microbes from the gut are then released in faecal material where they continue to decompose egested organic matter. Indeed, as mentioned above, earthworm casts contain different microbial populations to those in the parent material (Domínguez et al., 2010), and in turn it is expected that the inoculum of those communities in fresh organic matter promotes modifications similar to those found when earthworms are present, altering microbial community levels of activity and modifying the functional diversity of microbial populations in vermicomposting systems (Aira & Domínguez, 2011).

Upon completion of GAPs, the resultant earthworm casts undergo *cast associated processes* (CAPs; Figure 3), which are more closely related to ageing processes, the presence of unworked material and to physical modification of the egested material (weeks to months). During these processes the effects of earthworms are mainly indirect and derived from the GAPs (Aira et al., 2007). In addition, during this aging, vermicompost is expected to reach an optimum in terms of its biological properties, thereby promoting plant growth and suppressing plant diseases (Domínguez et al., 2010). However, little is yet known about when this "optimum" is achieved, how we can determine it in each case and if this "optimum" has some kind of expiration date.

4. Effects of earthworms on the structure and activity of microbial communities during vermicomposting

Since vermicomposting is a biological process, microorganisms play a key role in the evolution of the organic materials and in the transformations they suffer from wastes to safe organic amendments or fertilizers (vermicompost). Therefore, the effects that earthworms have on the microorganisms must be established because if the earthworms were to stimulate or depress microbiota or modify the structure and activity of microbial communities, they would have different effects on the decomposition of organic matter, and in turn on the quality of the final product. To address these questions we performed three laboratory experiments, with the following objectives:

i. To investigate whether and to what extent the earthworm *E. andrei* is capable of altering the structure and activity of microbial communities through the gut associated processes.
ii. To investigate how the earthworm species affect the structure and activity of microbial communities during the active phase of vermicomposting.
iii. To investigate the effectiveness of the active phase of vermicomposting for the short-term stabilization of a plant residue.

4.1 How do earthworms affect microbial communities through the gut associated processes?

To provide further light into the effect of gut transit on microbial communities, we carried out an experiment with microcosms filled with cow manure and inoculated with 25 mature individuals of the earthworm species *E. andrei*. The microcosms consisted of 250 mL plastic containers filled to three quarters of their capacity with sieved, moistened vermiculite. A plastic mesh was placed over the surface of the vermiculite and 100 g (fresh weight, fw) of the substrate was placed on top of the mesh, to avoid mixing the substrate with the vermiculite bedding (Figure 4a). The microcosms were covered with perforated lids and stored in random positions in an incubation chamber, at 20 °C and 90% relative humidity, for three days (Figure 4b). Control microcosms consisted of each type of manure incubated without earthworms. Each treatment was replicated five times. In order to obtain cast samples, earthworms were removed from the microcosms, washed three times with distilled water and placed in Petri dishes on moistened filter paper (Figure 4b). Casts from the same Petri dish were then collected with a sterile spatula and pooled for analysis in 1.5 mL Eppendorf tubes (Figure 4b); the same amount of manure samples were also collected from the control microcosms. Viable microbial biomass was determined as the sum of all identified phospholipid fatty acids (PLFAs) (Zelles, 1999). The structure of microbial communities was assessed by PLFA analysis; some specific PLFAs were used as biomarkers to determine the presence and abundance of specific microbial groups (Zelles, 1997). The sum of PLFAs characteristic of Gram-positive (iso/anteiso branched-chain PLFAs), and Gram-negative bacteria (monounsaturated and cyclopropyl PLFAs) were chosen to represent bacterial PLFAs, and the PLFA 18:2ω6c was used as a fungal biomarker. Total

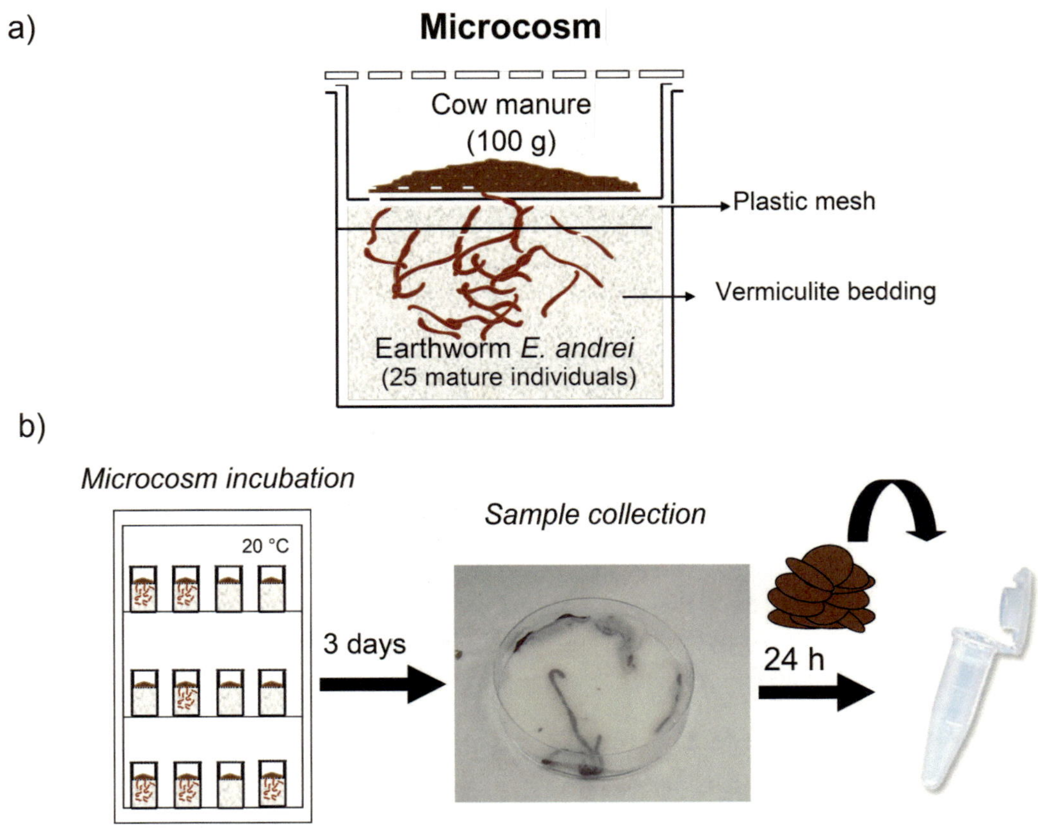

Fig. 4. Scheme of the (a) microcosm and (b) procedure for incubation of microcosms and collection of cast samples.

microbial activity was determined by hydrolysis of fluorescein diacetate (FDA), a colourless compound that is hydrolysed by both free and membrane bound enzymes, to release a coloured end product (fluorescein) that can be measured by spectrophotometry (Adam & Duncan, 2001). The data were analysed by a one-way ANOVA test, at $\alpha = 0.05$.

4.1.1 Microbial biomass

Recent reports suggest that the digestion of organic material by epigeic earthworms has negative effects on microbial biomass (Aira et al., 2006, 2009; Monroy et al., 2009). The present data are consistent with these findings, since we found a reduction in the viable microbial biomass as a result of the passage of the fresh substrate through the gut of the earthworm species *E. andrei* (Figure 5). More specifically, the total content of PLFAs was 1.5 times higher in the control treatment (1868.11 ± 129.02 µg g^{-1} dw) than that in earthworm casts (1249.87 ± 158.43 µg g^{-1} dw).

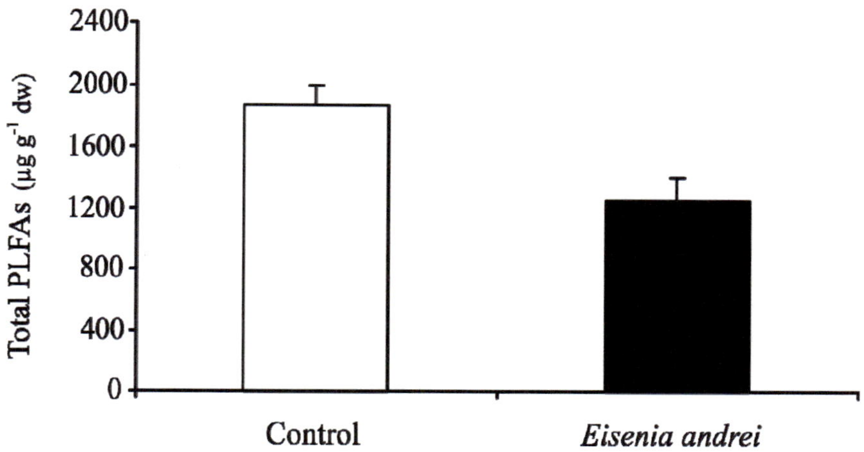

Fig. 5. Changes in the viable microbial biomass, measured as total PLFAs, after the passage of cow manure through the gut of the earthworm species *Eisenia andrei*. Values are means ± SE. Control is the manure incubated without earthworms.

Epigeic earthworms possess a diverse pool of digestive enzymes which enables them to digest bacteria, protozoa, fungi and partly decomposed plant debris (Zhang et al., 2000). Indeed, bacterial populations decreased in cow manure after transit through the earthworm gut (Figure 6a). As occurred with microbial biomass, bacterial PLFAs were 1.5 times lower in cast samples relative to the control (Figure 6a). However, the passage of cow manure through the earthworm gut affected fungal populations to a lesser extent than bacteria (Figure 6b).

Animal manures are microbial-rich environments in which bacteria constitute the largest fraction (around 70% of the total microbial biomass as assessed by PLFA analysis), with fungi mainly present as spores (Domínguez et al., 2010). Thus, earthworm activity is expected to have a greater effect on bacteria than on fungi in these organic substrates. These contrasting short-term effects on bacterial and fungal populations with earthworm activity are thus expected to have important implications on decomposition pathways during vermicomposting, because there exist important differences between both microbial

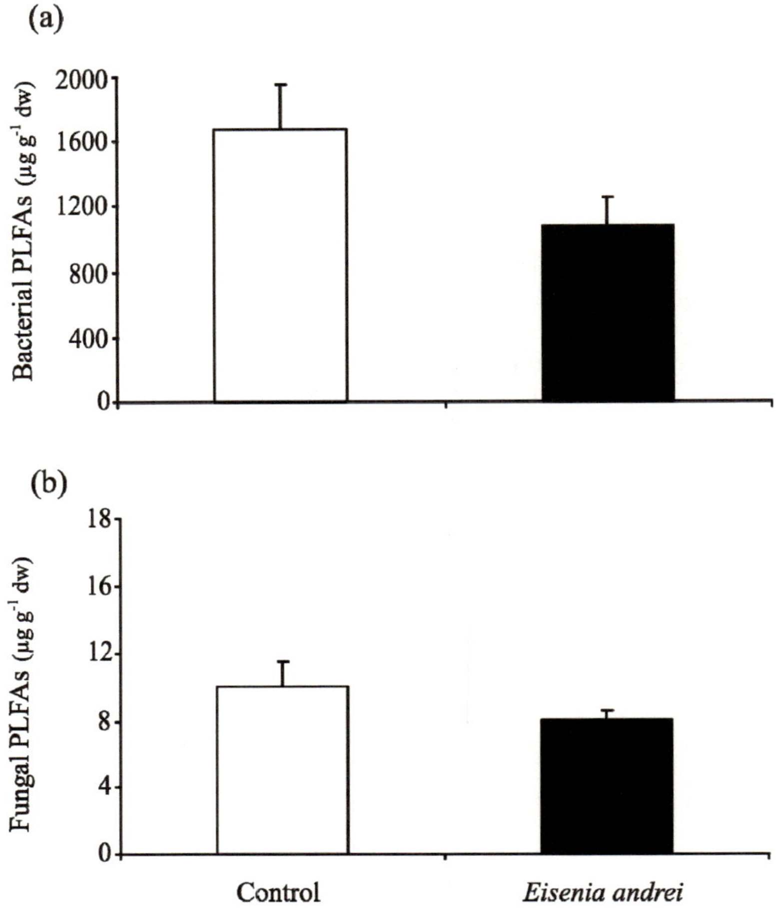

Fig. 6. Changes in (a) bacterial biomass calculated as the sum of the bacterial PLFA markers: i14:0, i15:0, a15:0, i16:0, a17:0, 16:1ω7, 17:1ω7, 18:1ω7, cy17:0 and cy19:0, and (b) PLFA 18:2ω6c, a measure of fungal biomass, after the passage of cow manure through the gut of the earthworm species *Eisenia andrei*. Values are means ± SE. Control is the manure incubated without earthworms.

decomposers related to resource requirements and exploitation. This is based on the fact that bacteria are more competitive in the use of readily decomposable compounds and have a more exploitative nutrient use strategy by rapidly using newly produced labile substrates (Bardgett & Wardle, 2010); whereas fungi are more competitive with regard to the degradation of more slowly decomposable compounds such as cellulose, hemicellulose and lignin (de Boer et al., 2005).

4.1.2 Microbial activity

The transit of the organic material through the gut of the earthworm *E. andrei* reduced the microbial activity, measured as FDA hydrolysis, relative to the control (Figure 7). We found up to a 30% reduction in the microbial activity from the control treatment (524.8 ± 60.1 μg fluorescein g^{-1} dw h^{-1}) to earthworm casts (208.0 ± 21.7 μg fluorescein g^{-1} dw h^{-1}). Similar decreases in microbial activity were reported in casts of *Eu. eugeniae* and *E. fetida* fed on pig and cow manures respectively (Aira et al., 2006; Aira & Domínguez, 2009).

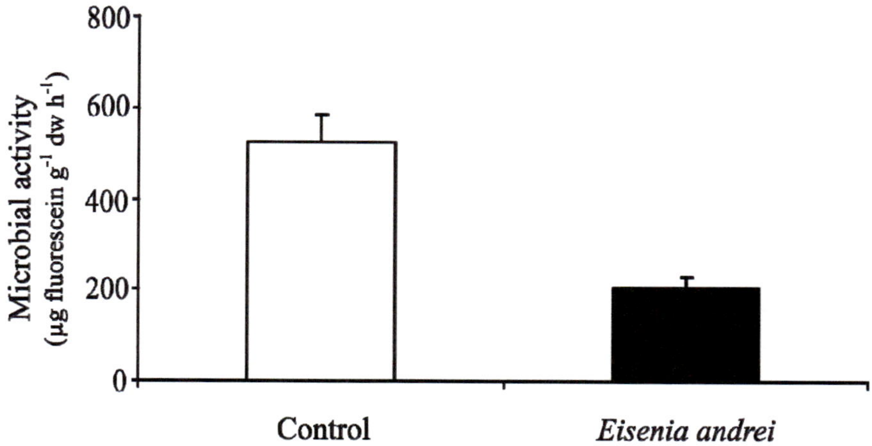

Fig. 7. Changes in microbial activity assessed by fluorescein diacetate hydrolysis, after the passage of cow manure through the gut of the earthworm species *Eisenia andrei*. Values are means ± SE. Control is the manure incubated without earthworms.

4.2 How does the earthworm species affect microbial communities?

Earthworms of different functional groups, or even different species within the same functional group, have a particular mode of food selection, ingestion, digestion, assimilation and movement, thus their importance in mixing, decomposition or nutrient release, as well as in the structure and activity of microbial communities will vary both qualitatively and quantitatively (Curry & Schmidt, 2007). To determine how the earthworm species shape the relationships between earthworms and microorganisms during the active phase of vermicomposting, we performed an experiment with mesocosms filled with cow manure and inoculated with 10 mature individuals of the earthworm species *Eisenia andrei*, *Eisenia fetida* and *Perionyx excavatus*. The mesocosms consisted of 2 L plastic containers filled to three quarters of their capacity with sieved, moistened vermiculite. A plastic mesh was placed over the surface of the vermiculite and 200 g (fresh weight, fw) of the substrate was placed on top of the mesh, to avoid mixing the substrate with the vermiculite bedding. The mesocosms were covered with perforated lids and stored in random positions in an incubation chamber, at 20 °C and 90% relative humidity. Control mesocosms consisted of each type of manure incubated without earthworms. Each treatment was replicated three times. The length of the active phase depends greatly on the rates at which the earthworms ingest and process the substrate (Domínguez et al., 2010). The high rate of consumption, digestion and assimilation of organic matter by these earthworm species resulted in the substrates being completely processed by the earthworms in one month, as previously shown by Lores et al. (2006). After this time (i.e., active phase), the earthworms were removed from the mesocosms and the processed material was collected from the surface of the vermiculite. The same amount of sample was also collected from the control mesocosms.

The viable microbial biomass was assessed as the sum of all identified PLFAs and certain PLFAs were used as biomarkers to determine the presence and abundance of specific microbial groups. Microbial community function was determined by measuring the bacterial and fungal growth rates. Bacterial growth was estimated by the incorporation of radioactively labelled leucine into proteins (Bååth, 1994), as modified by Bååth et al. (2001); fungal growth was estimated by the incorporation of radioactively labelled acetate into the fungal-specific lipid ergosterol Newell & Fallon (1991), with modifications by Bååth (2001). Total microbial activity was also assessed by measuring the rate of evolution of CO_2. The data were analyzed by a one-way ANOVA test. Post hoc comparisons of means were performed by a Tukey HSD test, at $\alpha = 0.05$.

4.2.1 Microbial biomass

The viable microbial biomass was about 3.8 times lower in the presence of *E. andrei* than that in the control (Figure 8), while no such pronounced decrease was detected in relation to the activity of *E. fetida* and *P. excavatus* (Figure 8). Similarly, the activity of *E. andrei* drastically reduced the bacterial and fungal biomass in cow manure, relative to the control (3.7 and 5.3 times, respectively), after the active phase of vermicomposting (Figure 9).

In the present study, the earthworm species *E. andrei* could have reduced the abundance of these microbial groups directly through ingestion, digestion and assimilation in the gut, and/or indirectly by accelerating the depletion of resources for the microbes, since greater losses of carbon were found as a result of earthworm activity after the active phase of vermicomposting (data not shown). However, the second explanation seems more likely to justify the reduction in fungal populations, since no significant changes were found in this microbial group after the passage through the gut of *E. andrei* (see experiment 1).

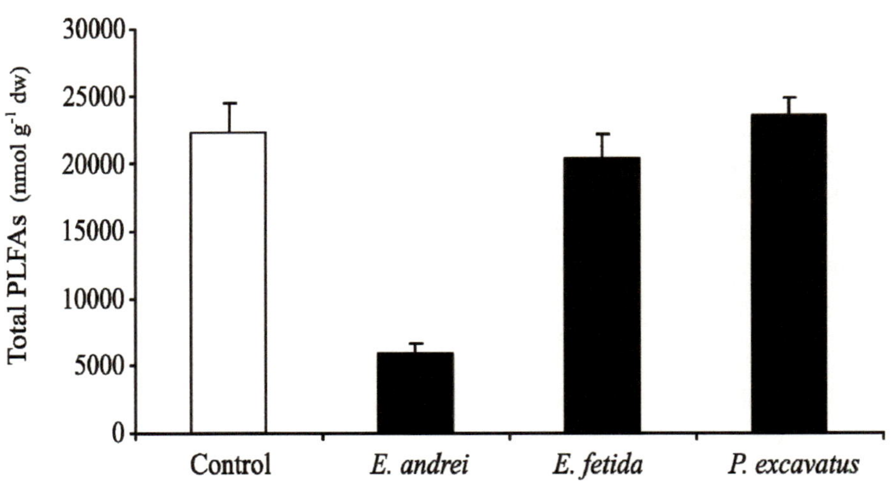

Fig. 8. Changes in the viable microbial biomass, measured as total PLFAs, of cow manure after being processed by the epigeic earthworm species *Eisenia andrei*, *Eisenia fetida* and *Perionyx excavatus* during the active phase of vermicomposting. Values are means ± SE. Control is the manure incubated without earthworms.

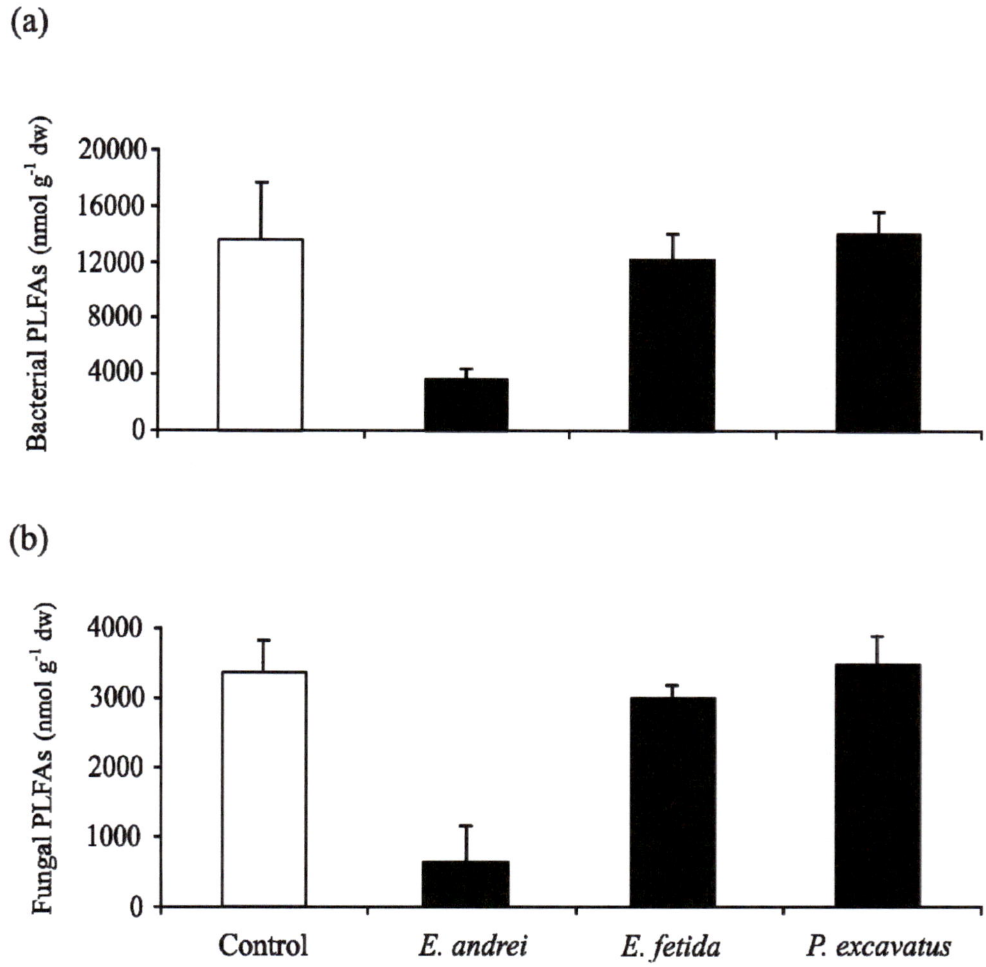

Fig. 9. Changes in (a) bacterial biomass calculated as the sum of the bacterial PLFA markers: i14:0, i15:0, a15:0, i16:0, i17:0, a17:0, 10Me16:0, 10Me17:0, 10Me18:0, 16:1ω7, 18:1ω7, cy17:0 and cy19:0, and (b) PLFA 18:2ω6c, a measure of fungal biomass, of cow manure after being processed by the epigeic earthworm species *Eisenia andrei*, *Eisenia fetida* and *Perionyx excavatus* during the active phase of vermicomposting. Values are means ± SE. Control is the manure incubated without earthworms.

4.2.2 Microbial activity

E. andrei reduced the bacterial growth rate by approximately 1.5 times relative to the control without earthworms after the active phase of vermicomposting (Figure 10a); no significant differences were detected with *E. fetida* and *P. excavatus* (Figure 10a). Despite the consistent effects on bacterial growth, earthworm activity did not affect the fungal growth rate (data not shown). Microbial activity in cow manure followed the same pattern as the bacterial growth rate (Figure 10b). As mentioned before, bacteria constitute the largest fraction of the microbiota in animal manures, and they are therefore expected to contribute greatly to the respiration rate.

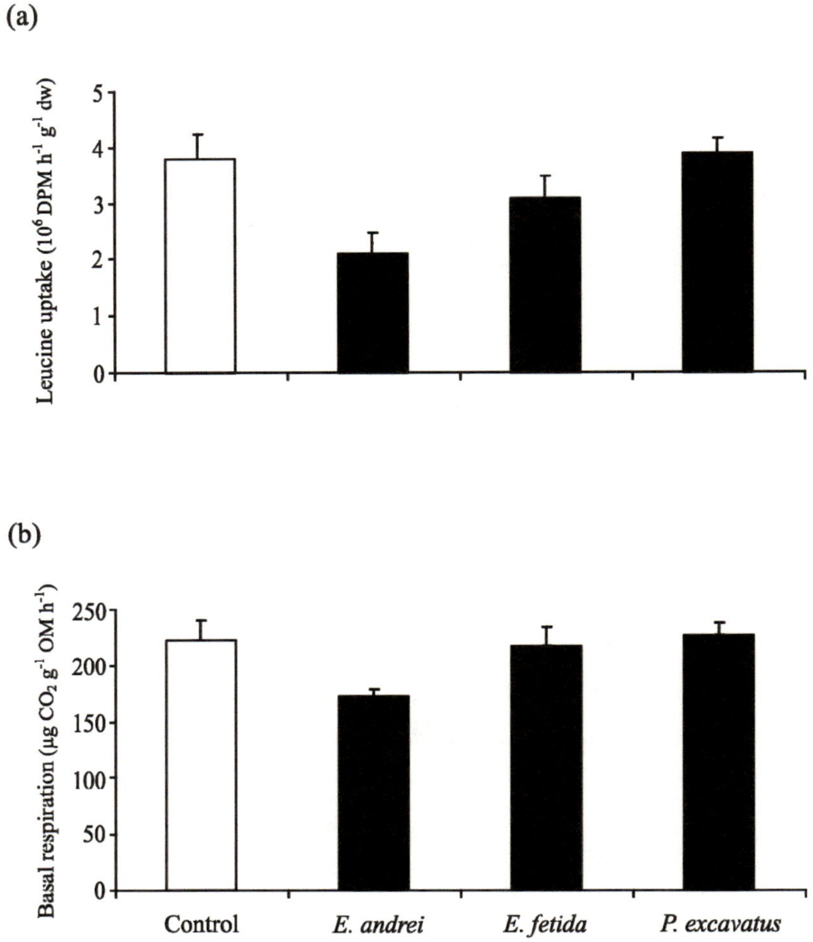

Fig. 10. Changes in (a) bacterial growth rate, estimated as leucine uptake and (b) microbial activity, measured as basal respiration, of cow manure after being processed by the epigeic earthworm species *Eisenia andrei, Eisenia fetida* and *Perionyx excavatus* during the active phase of vermicomposting. Values are means ± SE. Control is the manure incubated without earthworms.

The above-mentioned results highlight the potential of *E. andrei* for biodegrading organic substrates. The species *E. andrei* and *E. fetida* are closely related, although *E. andrei* predominates in mixed cultures, especially when there is no substrate limitation, as occurred in this experiment, indicating that it is a more extreme r strategist than *E. fetida*, as shown by more rapid growth and reproduction (Domínguez et al., 2005).

4.3 How do earthworms affect microbial communities of a plant residue in the short-term?

In this study we evaluated the effectiveness of the active phase of vermicomposting for the short-term stabilization of grape marc, a lignocellulosic enriched residue that consists of the stalks, skin, pulp and seeds remaining after the grape crushing and pressing stages in wine production (Flavel et al., 2005). This by-product is a valuable resource as a soil fertilizer with high contents of macro- and micronutrients for crop growth (Bertran et al., 2004). However, the overproduction of grape marc - more than 750,000 ton per year in Spain (Fernández-Bayo et al., 2007) - has become a problem that requires strategies for its disposal and/or management. Whilst composting has been widely used for the treatment of winery wastes (Bertran et al., 2004; Marhuenda-Egea et al., 2007; Fernández et al., 2008; Bustamante et al., 2009; Paradelo et al., 2010), there are still very few studies on the application of vermicomposting as a methodological alternative to recycling such wastes (Nogales et al., 2005; Romero et al., 2007, 2010).

The vermicomposting of grape marc was performed in mesocosms that consisted of plastic containers (2 L), which were filled to three quarters of the capacity with moistened (80% moisture content) and mature vermicompost in order to ensure the survival of the earthworms. Five hundred juvenile and adult specimens of the epigeic earthworm species *Eisenia andrei* were placed on the surface of the vermicompost. One kilogram (fresh weight) of grape marc was placed on a mesh on the surface of the vermicompost and was rewetted by spraying it with 20mL of tap water. The mesocosms were covered with perforated lids and stored in random positions in an incubation chamber, at 20 °C and 90% relative humidity. Control mesocosms consisted of the grape marc incubated without earthworms. Each treatment was replicated five times. The high density of earthworms used and the relatively rapid gut transit time of the epigeic earthworm species *E. andrei*, around 2.5–7 h, resulted in the grape marc being completely processed by the earthworms in 15 days. After this time (i.e., active phase), the earthworms were removed from the mesocosms and the processed material was collected from the surface of the vermicompost bedding. The same amount of sample was also collected from the control mesocosms. The viable microbial biomass was assessed as the sum of all identified PLFAs and certain PLFAs were used as biomarkers to determine the presence and abundance of specific microbial groups. Microbial community function was determined by measuring the total microbial activity assessed by basal respiration, and by determining the activity of enzymes involved in C and N cycles, i.e. protease and cellulase activities.

4.3.1 Microbial biomass

Earthworm activity reduced the viable microbial biomass measured as total PLFAs relative to the control without earthworms (96.90 ± 1.04 μg mL $^{-1}$ and 113.60 ± 1.04 μg mL $^{-1}$ for treatments with and without earthworms). Similarly, the presence of earthworms also reduced the abundance of both bacteria and fungi after the active phase of vermicomposting of grape marc (Figure 11).

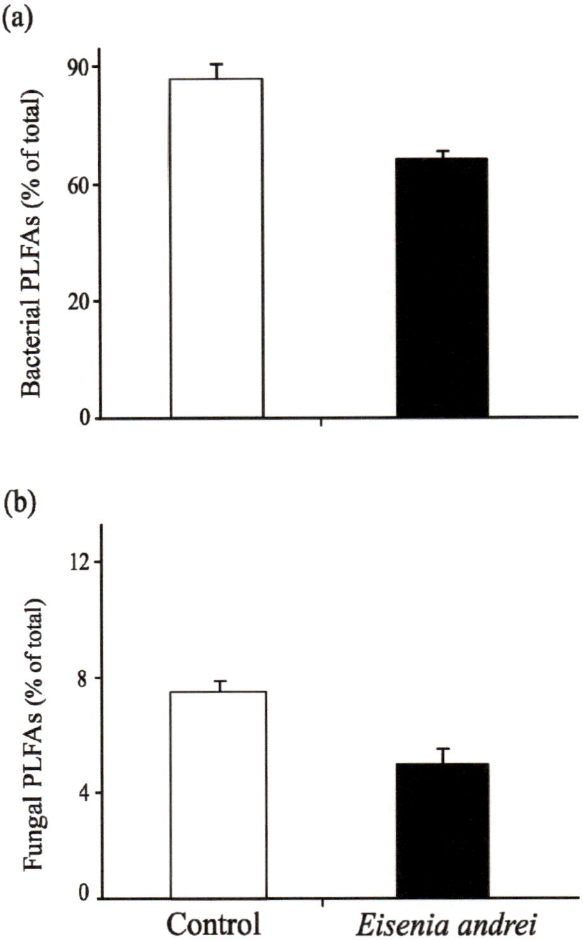

Fig. 11. Changes in (a) bacterial biomass calculated as the sum of the bacterial PLFA markers: i14:0, i15:0, a15:0, i16:0, i17:0, a17:0, 16:1ω7, 17:1ω7, cy17:0 and cy19:0, and (b) PLFAs 18:1ω9c and 18:2ω6c, a measure of fungal biomass, of grape marc after being processed by the epigeic earthworm species *Eisenia andrei* during the active phase of vermicomposting. Values are means ± SE. Control is the grape marc incubated without earthworms.

4.3.2 Microbial activity

As occurred in the two previous experiments, the total microbial activity measured as basal respiration was about 1.7 times lower in the presence of *E. andrei* than that in the control without earthworms (Figure 12). This suggests that the presence of earthworms favoured the stabilization of the residue, as shown by Lazcano et al. (2008). These authors found that both vermicomposting treatments (vermicomposting and a combination of composting and vermicomposting) produced more stabilized substrates than the active phase of composting in terms of microbial activity.

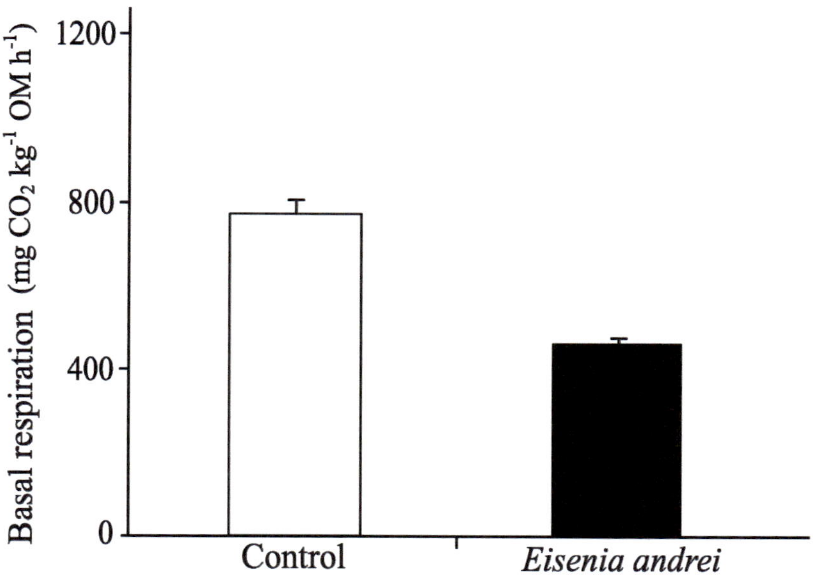

Fig. 12. Changes in microbial activity assessed by basal respiration of grape marc after being processed by the epigeic earthworm species *Eisenia andrei* during the active phase of vermicomposting. Values are means ± SE. Control is the grape marc incubated without earthworms.

The study of enzyme activities has been shown to be a reliable tool for characterizing the state and evolution of the organic matter during vermicomposting (Benítez et al., 2005), as they are implicated in the biological and biochemical processes that transform organic wastes into stabilized products. In the present study, earthworm activity greatly reduced the activities of the protease (Figure 13a) and cellulase enzymes (Figure 13b) in comparison with the control. These findings are in agreement with microbial activity data, which reinforces that a higher degree of stability was reached after the active phase of vermicomposting. Similarly, Lazcano et al. (2008) reported lower values of protease activity, relative to the control, after vermicomposting and composting with subsequent vermicomposting (3 and 4.4 times lower, respectively). However, they did not find any differences in relation to this enzyme activity after the active phase of composting, indicating that the vermicomposted materials were significantly more stabilized than the compost.

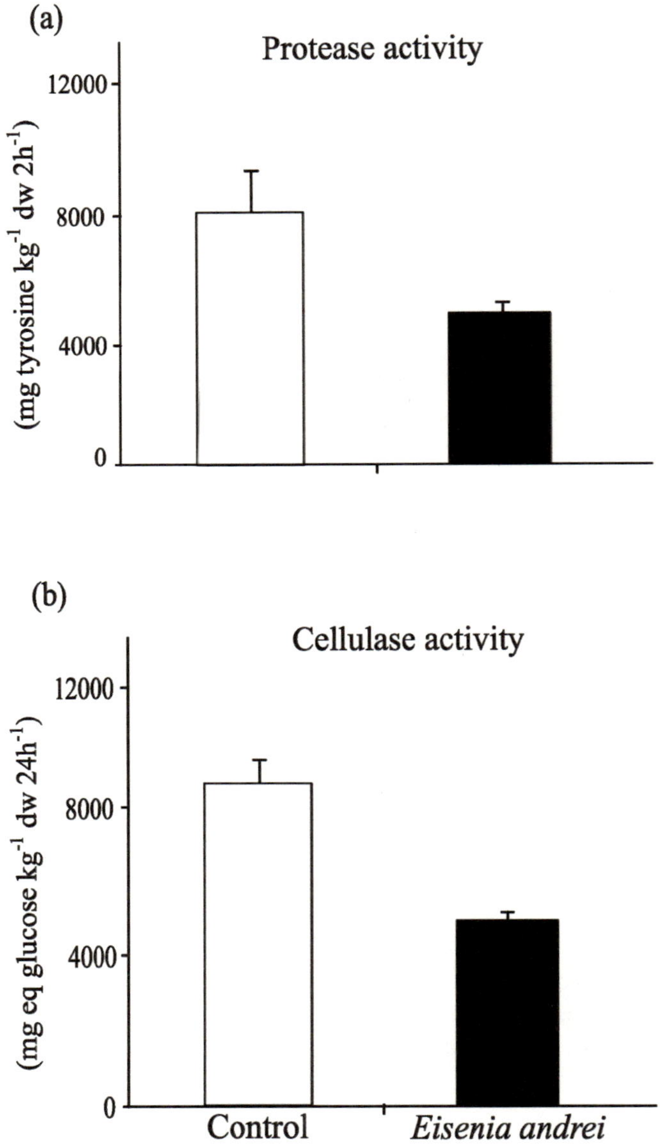

Fig. 13. Changes in (a) protease activity, and (b) cellulase activity of grape marc after being processed by the epigeic earthworm species *Eisenia andrei* during the active phase of vermicomposting. Values are means ± SE. Control is the grape marc incubated without earthworms.

5. Conclusions

Detritivorous earthworms interact intensively with microorganisms during vermicomposting, thus accelerating the stabilization of organic matter and greatly modifying its physical and biochemical properties. Digestion of the ingested material is the first step in earthworm-microorganism interactions. Passage of organic material through the gut of epigeic earthworms reduced the viable microbial biomass and affected the abundance of bacteria to a greater extent than fungi. Microbial activity also decreased after transit of the microorganisms through the earthworm gut. Accordingly, the presence of earthworms reduced microbial biomass and activity after the active phase of vermicomposting, although this effect depended on the earthworm species involved. The bacterial growth rate also decreased in the substrate, whereas the fungal growth rate was not affected after one month. The speed at which these transformations occurred made the active phase of vermicomposting a suitable stage for studying the relationships between earthworms and microorganisms and permitted us to understand the chemical and biological consequences of earthworm activities. Ultimately, these findings provide valuable information for the understanding of the transformations that organic matter undergoes during vermicomposting and, in addition constitute a powerful tool for the development of strategies leading to a more efficient process for the disposal and/or management of organic wastes.

6. Acknowledgments

This research was financially supported by the Spanish Ministerio de Ciencia e Innovación (CTM2009-08477). María Gómez Brandón is financially supported by a postdoctoral research grant from Fundación Alfonso Martín Escudero.

7. References

Adam, G. & Duncan, H. (2001). Development of a Sensitive and Rapid Method for the Measurement of Total Microbial Activity Using Fluorescein Diacetate (FDA) in a Range of Soils. *Soil Biology and Biochemistry*, vol. 33, No. 7-8, pp. 943-951, ISSN 0038-0717.

Aira, M., Monroy, F. & Domínguez, J. (2006). Changes in Microbial Biomass and Microbial Activity of Pig Slurry After the Transit Through the Gut of the Earthworm *Eudrilus eugeniae* (Kinberg, 1867). *Biology and Fertility of Soils*, vol. 42, No. 4, pp. 371–376, ISSN 0178-2762.

Aira, M., Monroy, F. & Domínguez, J. (2007). *Eisenia fetida* (Oligochaeta: Lumbricidae) Modifies the Structure and Physiological Capabilities of Microbial Communities Improving Carbon Mineralization During Vermicomposting of Pig Manure. *Microbial Ecology*, vol. 54, No. 4, pp. 662-671, ISSN 0095-3628.

Aira, M. & Domínguez, J. (2008). Optimizing Vermicomposting of Animal Wastes: Effects of Dose of Manure Application on Carbon Loss and Microbial Stabilization. *Journal of Environmental Management*, vol. 88, No. 4, pp. 1525-1529, ISSN 0301-4797.

Aira, M.; Sampedro, L.; Monroy, F. & Domínguez, J. (2008). Detritivorous Earthworms Directly Modify the Structure, thus Altering the Functioning of a Microdecomposer Food Web. *Soil Biology and Biochemistry*, vol. 40, No. 10, pp. 2511-2516, ISSN 0038-0717.

Aira, M. & Domínguez, J. (2009). Microbial and Nutrient Stabilization of Two Animal Manures After the Transit Through the Gut of the Earthworm *Eisenia fetida* (Savigny, 1826). *Journal of Hazardous Materials*, vol. 161, No. 2-3, pp. 1234-1238, ISSN 0304-3894.

Aira, M.; Monroy, F. & Domínguez, J. (2009). Changes in Bacterial Numbers and Microbial Activity of Pig Slurry during Gut Transit of Epigeic and Anecic Earthworms. *Journal of Hazardous Materials*, vol. 162, No. 2-3, pp. 1404-1407, ISSN 0304-3894.

Aira, M. & Domínguez, J. (2011). Earthworm Effects without Earthworms: Inoculation of Raw Organic Matter with Worm-Worked Substrates Alters Microbial Community Functioning. *Plos One*, vol. 6, No. 1, pp. 1-8, ISSN 1932-6203.

Bååth, E. (1994). Measurement of Protein Synthesis by Soil Bacterial Assemblages with the Leucine Incorporation Technique. *Biology and Fertility of Soils*, vol. 17, No. 2, pp. 147–153, ISSN 0178-2762.

Bååth, E. (2001). Estimation of Fungal Growth Rates in Soil Using ^{14}C-Acetate Incorporation into Ergosterol. *Soil Biology and Biochemistry*, vol. 33, No. 14, pp. 2011–2018, ISSN 0038-0717.

Bååth, E.; Petterson, M. & Söderberg, K.H. (2001). Adaptation of a Rapid and Economical Microcentrifugation Method to Measure Thymidine and Leucine Incorporation by Soil Bacteria. *Soil Biology and Biochemistry*, vol. 33, No. 11, pp. 1571-1574, ISSN 0038-0717.

Bardgett, R.A. & Wardle, D.A. (2010). *Aboveground-Belowground Linkages*, Oxford University Press, ISBN 978-0-19-954687-9, New York.

Benítez, E.; Sanz,H. & Nogales, R. (2005). Hydrolytic Enzyme Activities of Extracted Humic Substances During the Vermicomposting of a Lignocellulosic Olive Waste, *Bioresource Technology* vol. 96, No. 7, pp. 785–90, ISSN 0960-8524.

Bertran, E.; Sort, X.; Soliva, M. & Trillas, I. (2004). Composting of Winery Waste: Sledges and Grape Stalks. *Bioresource Technology*, vol. 95, No. 2, pp. 203–208, ISSN 0960-8524.

Bustamante, M.A.; Paredes, C.; Morales, J.; Mayoral, A.M. & Moral, R. (2009). Study of the Composting Process of Winery and Distillery Wastes Using Multivariate Techniques. *Bioresource Technology*, vol. 100, No. 20, pp. 4766–4772, ISSN 0960-8524.

Curry, J.P. & Schmidt, O. (2007). The Feeding Ecology of Earthworms - a Review. *Pedobiologia*, vol. 50, No. 6, pp. 463–477, ISSN 0031-4056.

Darwin, C . (1881) *The Formation of Vegetable Mould through the Action of Worms with Observations on their Habits*, Murray, London.

De Boer, W.; Folman, L.B.; Summerbell, R.C. & Boddy, L. (2005). Living in a Fungal World: Impact of Fungi on Soil Bacterial Niche Development. *FEMS Microbiology Reviews*, vol. 29, No. 4, pp. 795-811, ISSN 1574-6976.

Domínguez, J. (2004). State of the Art and New Perspectives on Vermicomposting Research, In: *Earthworm Ecology*, C.A. Edwards, (Ed.), 401-424, CRC Press, ISBN 1-884015-74-3, Boca Raton, Florida.

Domínguez, J.; Ferreiro, A. & Velando, A. (2005). Are *Eisenia fetida* (Savigny, 1826) and *Eisenia andrei* Bouché, 1972 (Oligochaeta, Lumbricidae) Different Biological Species? *Pedobiologia*, vol. 49, pp. 81-87, ISSN 0031-4056.

Domínguez, J. & Edwards, C.A. (2010a). Relationships between Composting and Vermicomposting: Relative Values of the Products, In: *Vermiculture Technology: Earthworms, Organic Waste and Environmental Management*, C.A. Edwards; N.Q. Arancon; R.L. Sherman, (Eds.), 1-14, CRC Press, ISBN 9781439809877, Boca Raton, Florida.

Domínguez, J. & Edwards, C.A. (2010b). Biology and Ecology of Earthworm Species Used for Vermicomposting, In: *Vermiculture Technology: Earthworms, Organic Waste and*

Environmental Management, C.A. Edwards; N.Q. Arancon; R.L. Sherman, (Eds.), 25-37, CRC Press, ISBN 9781439809877, Boca Raton, Florida.

Domínguez, J.; Aira, M. & Gómez-Brandón, M. (2010). Vermicomposting: Earthworms Enhance the Work of Microbes, In: *Microbes At Work: From Wastes to Resources*, H. Insam; I. Franke-Whittle; M. Goberna, (Eds.), 93-114, Springer, ISBN 978-3-642-04042-9, Heidelberg, Germany.

Drake, H.L. & Horn, M.A. (2007). As the Worm Turns: the Earthworm Gut as a Transient Habitat for Soil Microbial Biomes. *Annual Review of Microbiology*, vol. 61, pp. 169-189, ISSN 0066-4227.

Edwards, C.A. & Bohlen, P.J. (1996). *Biology and Ecology of Earthworms*, Chapman and Hall, London.

Flavel, T.C., Murphy, D.V., Lalor, B.M., & Fillery, I.R.P. (2005). Gross N Mineralization Rates after Application of Composted Grape Marc of Soil. *Soil Biology and Biochemistry*, vol. 37, No. 7, pp. 1397–1400, ISSN 0038-0717.

Fernández-Bayo, J.D.; Nogales, R. & Romero, E. (2007). Improved Retention of Imidacloprid (Confidor®) in Soils by Adding Vermicompost from Spent Grape Marc. *Science of Total Environment*, vol. 378, No. 1-2, pp. 95–100, ISSN 0048-9697.

Fernández, F.J.; Sánchez-Arias, V.; Villaseñor, J. & Rodríguez, L. (2008). Evaluation of Carbon Degradation during Co-composting of Exhausted Grape Marc with Different Biowastes. *Chemosphere*, vol. 73, No. 5, pp. 670-677, ISSN 0045-6535.

Gómez-Brandón, M.; Aira, M.; Lores, M. & Domínguez, J. (2011a). Epigeic Earthworms Exert a Bottleneck Effect on Microbial Communities through Gut Associated Processes. *Plos One*, vol. 6, No. 9, pp.1-9, ISSN 1932-6203.

Gómez-Brandón, M.; Aira, M.; Lores, M. & Domínguez, J. (2011b). Changes in Microbial Community Structure and Function During Vermicomposting of Pig Slurry. *Bioresource Technology*, vol. 102, No. 5, pp. 4171-4178, ISSN 0960-8524.

Hutchinson, M.L.; Walters, L.D.; Avery, S.M.; Munro, F. & Moore, A. (2005). Analyses of Livestock Production, Waste Storage, and Pathogen Levels and Prevalences in Farm Manures. *Applied and Environmental Microbiology*, vol. 71, No. 3, pp. 1231-1236, ISSN 1098-5336.

Lazcano, C.; Gómez-Brandón, M. & Domínguez, J. (2008). Comparison of the Effectiveness of Composting and Vermicomposting for the Biological Stabilization of Cattle Manure. *Chemosphere*, vol. 72, No. 7, pp. 1013-1019, ISSN 0045-6535.

Lores, M.; Gómez-Brandón, M.; Pérez-Díaz, D. & Domínguez, J. (2006). Using FAME profiles for the Characterization of Animal Wastes and Vermicomposts. *Soil Biology and Biochemistry*, vol. 38, No. 9, pp. 2993-2996, ISSN 0038-0717.

Marhuenda-Egea, F.C.; Martínez-Sabater, E.; Jordá, J.; Moral, R.; Bustamante, M.A.; Paredes, C. & Pérez-Murcia, M.D. (2007). Dissolved Organic Matter Fractions Formed during Composting of Winery and Distillery Residues: Evaluation of the Process by Fluorescence Excitation–Emission Matrix. *Chemosphere*, vol. 68, No. 2 pp. 301–309, ISSN 0045-6535.

Monroy, F.; Aira, M. & Domínguez, J. (2008). Changes in Density of Nematodes, Protozoa and Total Coliforms after Transit through the Gut of Four Epigeic Earthworms (Oligochaeta). *Applied Soil Ecology*, vol. 39, No. 2, pp. 127–132, ISSN 0929-1393.

Monroy, F.; Aira, M. & Domínguez, J. (2009). Reduction of Total Coliform Numbers during Vermicomposting is Caused by Short-term Direct Effects of Earthworms on

Microorganisms and Depends on the Dose of Application of Pig Slurry. *Science of the Total Environment*, vol. 407, No. 20, pp. 5411-5416, ISSN 0048-9697.

Monroy, F.; Aira, M. & Domínguez, J. (2011). Epigeic Earthworms Increase Soil Arthropod Populations during First Steps of Decomposition of Organic Matter. *Pedobiologia*, vol. 54, No. 2, pp. 93-99, ISSN 0031-4056.

Moore, J.C.; Berlow, E.L.; Coleman, D.C.; de Ruiter, P.C.; Dong, Q.; Johnson, N.C.; McCann, K.S.; Melville, K.; Morin, P.J.; Nadelhoffer, K.; Rosemond, A.D.; Post, D.M.; Sabo, J.L.; Scow, K.M.; Vanni, M.J. & Wall, D.H. (2004). Detritus, Trophic Dynamics and Biodiversity. *Ecology Letters*, vol. 7, pp. 584–600.

Moral, R.; Paredes, C.; Bustamante, M.A.; Marhuenda-Egea, R. & Bernal, M.P. (2009). Utilisation of Manure Composts by High-Value Crops: Safety and Environmental Challenges. *Bioresource Technology*, vol. 100, No. 22, pp. 5454-5460, ISSN0960-8524.

Newell, S.Y. & Fallon, R.D. (1991). Toward a Method for Measuring Instantaneous Fungal Growth Rates in Field Samples. *Ecology*, vol. 72, No. 5, pp. 1547–1559, ISSN 0012-9658.

Nogales, R.; Cifuentes, C. & Benítez E. (2005). Vermicomposting of Winery Wastes: a Laboratory Study. *Journal of Environmental Science and Health Part B* vol. 40, No. 4, pp. 659-673, ISSN 0360-1234.

Paradelo, R.; Moldes, A. & Barral, M. (2010). Utilization of a Factorial Design to Study the Composting of Hydrolyzed Grape Marc and Vinification Lees. *Journal of Agriculture and Food Chemistry*, vol. 58, No. 5, pp. 3085–3092.

Reynolds, J. W. & Wetzel, M. J. (2010). *Nomenclatura Oligochaetologica. Supplementum Quartum. A catalogue of names, descriptions and type specimens of the Oligochaeta*, Illinois Natural History Survey Special Publication, Chicago.

Romero, E.; Plaza, C.; Senesi, N.; Nogales, R. & Polo, A. (2007). Humic Acid-Like Fractions in Raw and Vermicomposted Winery and Distillery Wastes. *Geoderma*, vol. 139, No. 3-4, pp. 397-406, ISSN 0016-7061.

Romero, E.; Fernández-Bayo, J; Díaz, J.M.C. & Nogales, R. (2010). Enzyme Activities and Diuron Persistence in Soil Amended with Vermicompost Derived from Spent Grape Marc and Treated with Urea. *Applied Soil Ecology*, vol. 44, No. 3, pp.198-204, ISSN 0929-1393.

Schönholzer, F.; Dittmar, H. & Zeyer, J. (1999). Origins and Fate of Fungi and Bacteria in the Gut of *Lumbricus terrestris* L. Studied by Image Analysis. *FEMS Microbiology Ecology*, vol. 28, No. 3, pp. 235-248, ISSN 1574-6941.

Zelles, L. (1997). Phospholipid Fatty Acid Profiles in Selected Members for Soil Microbial Communities. *Chemosphere*, No. 1-2, vol. 35, pp. 275–294, ISSN 0045-6535.

Zelles, L. (1999). Fatty Acid Patterns of Phospholipids and Lipopolysaccharides in the Characterization of Microbial Communities in Soil: a Review. *Biology and Fertility of Soils*, vol. 29, No. 2, pp. 111–129, ISSN 0960-8524.

Zhang, B.; Li, G.; Shen, T.; Wang, J. & Sun, Z. (2000). Changes in Microbial Biomass C, N, and P and Enzyme Activities in Soil Incubated with the Earthworms *Metaphire guillelmi* or *Eisenia fetida*. *Soil Biology and Biochemistry*, vol. 32, No. 2, pp. 2055-2062, ISSN 0038-0717.

3

The Sanitation of Animal Waste Using Anaerobic Stabilization

Ingrid Papajová and Peter Juriš
Institute of Parasitology of the Slovak Academy of Sciences
Slovak Republic

1. Introduction

Animal production pose potential hazards of environmental contamination with pathogenic microorganisms. These are particularly related to a subsequent storage processing and utilization of animal organic wastes (manure, fertilizer, wastewater, sludges etc.). A major source of pathogenic microorganisms in the environment are excrements from clinically and subclinically infected farm animals.

Handling, storage, treatment and use of different forms of animals excrements entails two principle problems: epizootological or epidemiological and hygienical. Solid excrements contain high numbers of common intestinal microflora (*E. coli*, faecal streptococci, lactobacilli etc.), bacteria that are pathogenic also for man (salmonellae, mycobacteria, listeriae etc.), protozoa (*Isospora* spp., *Balantidium coli*) and eggs or larvae of enteronematodes (*Ascaris suum, Oesophagostomum* sp., *Trichuris suis* etc.) (Lauková et al., 2000; Krupicer et al., 2000).

The parasitic propagative stages, mainly endoparasitic protozoa and helminths develop mostly outside their host´s organism. *A. suum* eggs are hygienically the most problematic ones. They are amongst the helminth eggs most resistant to environmental factors and may survive in the nature for many years, therefore, they tend to accumulate in the environment (soil, water) and serve as an infectious entity for both man and animals (Papajová and Juriš, 2009). The cell wall of *A suum* egg is enveloped with an outer layer formed by acid polysaccharides and proteins, central layer consisting of proteins (25%) and lipids (75%, particularly alpha glycosides). Thus this resistant cell wall protects eggs against effects of chemicals and drying (Eckert, 1992).

Regarding the spread of helminthoses, domestic animals (dogs, cats) are also of great importance because they live in a close contact with man. Infection and way of transmission of the disease depends on the way of breeding and on the breeding environment where the animal occurs. An important factor of the risk of infection transmission is also possibility of animal to move outside its housing (yard, move in nature), or the use of a dog (hunting or social). The most frequent way of transmission of parasitic diseases is through the contact (free-living animals with a domestic ones), or through contaminated environment with developmental stages (oocysts, eggs, larvae). Through faeces of infected dogs and cats the germs of parasitozoonoses spread into the environment. It is especially the case of cysts of intestinal parasitic protozoa - *Entamoeba histolytica, Giardia intestinalis, Balantidium coli, Toxoplasma gondii,* the eggs of tapeworms *Dipylidium* sp., *Echinococcus* sp. and parasitic nematodes (Antolová et al., 2004; Matsuo and Nakashio, 2005; Miterpáková et al., 2006). Regarding public health helminthozoonoses caused by *Toxocara* sp. (in dogs) and *Ascaris* sp. (in pigs) are very significant, especially due to their zoonotic character connected with the syndrome *larva migrans*.

Humans became infected usually orally *(per os)* by ingestion of substrates (soil, vegetables, etc.) with embryonated *Toxocara* eggs. Many symptoms are associated with this infection, including changes in blood cell counts and affection of various organs, as the ascaris larvae can migrate throughout the body. The symptoms of infection are often non-specific and may be mistaken with those of other infectious agents (common viral diseases, diarrhoea) or we may not observe any clinical signs. Toxocariasis manifests itself in two distinct forms: visceral, *larva migrans visceralis,* and ocular, *larva migrans ocularis* (Despommier, 2003).

A. suum infects pigs and is of major economical significance due to production losses linked to reduced feed conversion efficiency and losses to the mean industry associated with the condemnation of "milk-spot" livers (Dubinský et al., 2000). *Ascaris* infects over a quarter of the world´s human population (1.47 billion people worldwide) and clinically affects ~335 million people (Crompton, 1999).

The above-mentioned helminthozoonoses are classified among epidemiologically "low-risk" parasitozoonoses, because the propagative stages develop in the outdoor environment into the infectious stage and potentially secondarily contaminate the food chain. Therefore, a direct contact with infected animal, but also contaminated environment, or contaminated food chain (water, vegetables) are considered as a potential risk factor.

Attention has receantly been paid also to the problem of hygienical hazards in terms of the treatment and use of animal excrements and their application to soil as valuable nutrients for cultivated plants. The hazards are mainly connected with the quantity of continually produced solid and liquid wastes. The results is ecological disbalance, mainly with respect to environmental load with pathogenic microorgamism and nitrous organic substances. Animal organic wastes are also sources of greate amounts of gases releases. The most dangerous of them are ammonia and methane. Ammonia released into the atmosphere is irritating and toxic to the biotic component of the environment. On the other hand, animal excrements can supply essential plant nutrients and improve the fertility of soil by adding organic matter.

Therefore, to prevent health risks (for human as well as for animals) and odour nuisance from animal wastes, different methods for a satisfactory utilisation and sanitation have been researched (Schwartzbrod et al., 1989, Tofant et al., 1999; Juriš et al., 2000; Sasáková et al., 2005; Papajová and Juriš, 2009). There are big variations in the treatment of animal wastes (aerobic and anaerobic stabilization, composting etc.).

For the above-mentioned reasons our studies concentrated on anaerobic stabilization of liquid (slurry) and solid (manure, excrements) animal waste. The aims of our study were to monitor the physical-chemical changes in pig slurry treated by ecologically acceptable and energetically beneficial anaerobic stabilisation, changes in the properties of anaerobically digested slurry stored in ground lagoon and the effect of anaerobically digested slurry stores in ground lagoon on the survival of parasitic germs. The impact of lime (two types) on the survival of parasitic organisms in anaerobic stored manure and dog excrements mixed with hay, was also studied.

2. Materials and methods

2.1 Parasitological methods

To determine helminth eggs count in slurry (input and output samples from bioreactor and in lagoon samples - supernatant and sediment), 50 ml from each of the 1 l sample was taken and examined by a sedimentation-flotation mathod (Cherepanov, 1982).

A. suum eggs were isolated by dissection of a distal uterine part of female pig ascaris. The distal uterine ends were then removed to a glass homogenizer and processed. The water suspension of eggs was stored in an Erlenmayer flask in a refrigerator at 4°C.

We used the "artificial contamination of lagoon and organic wastes" approach to make sure that there is s sufficient number of positive samples in our observations.

Model eggs were inoculated by a micropipette into polyurethane carriers, prepared according to Plachý and Juriš (1995), at a dose of 1 000 eggs per one carrier. A porous cellular plastic - soft expanded polyurethane, commercially known as a plastic foam, was used as a material for the carriers. It is an additive product of polyisocyanates and compounds with a high content of hydroxylic groups. It consists of a network of interconnected cells, resembling a honeycomb. Its polyurethane structure allows for a sufficient contact of helminth eggs with the environment, preventing them from release and consequently improving their recovery (Picture 1). For mechanical protection the carriers were placed to perforated plastic net (Picture 2) before introducing them into the organic wastes. Three samples were taken and analysed at each sampling interval. The eggs were re-isolated from the inoculated carriers as follows: the carriers were cut into small pieces and washed in a mortar with 3 · 5 ml portions of saline, thoroughly stirred and filtered through a sieve into test tubes. After centrifugation, sediments were transferred to Petri dishes.

The viability of exposed unembryonated *A. suum* eggs was determined by incubation up to the embryonated stage (Picture 6) in a thermostat at 26°C for 21 days. Petri dishes with *A. suum* eggs were aerated daily with micropipette. The developmental ability of *A. suum* eggs was compared with that of the control eggs which were kept in distilled water under aerobic conditions.

2.2 Physical and chemical methods

The following changes in physical and chemical properties of the solid and liquid wastes were monitored: pH, dry mater (DM), inorganic (IM) and organic (OM) matter, ammonium ions (NH_4^+), total nitrogen (N_t), chemical oxygen demand (COD), soluble and insoluble substances and C:N ratio.

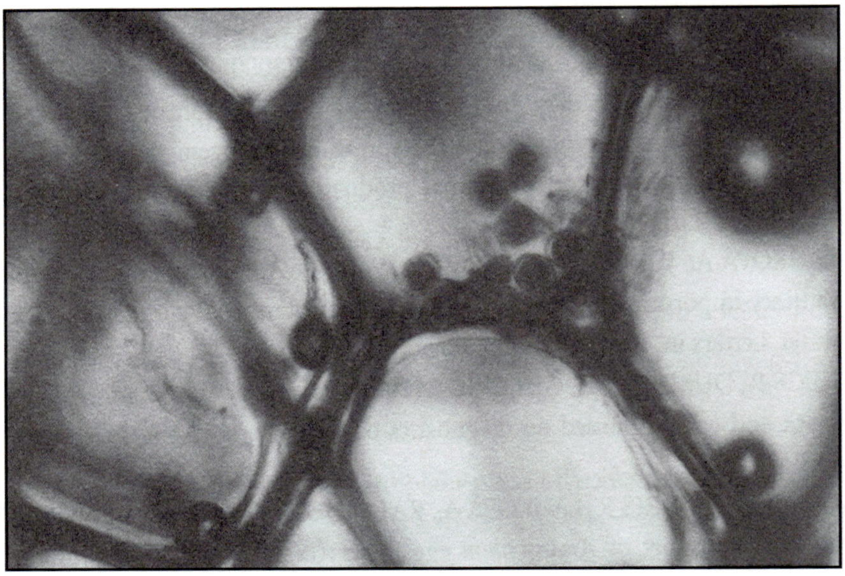

Picture 1. Polyurethane carrier with adhered *A. suum* eggs

Picture 2. Perforated plastic nets with the carriers

The samples were examined for the pH (1:10 water extract) using a pH electrode (HACH Company, Loveland, Colorado, USA). Dry matter (drying at 105°C to a constant weight), residum-on-ignition (550°C for 4 h), and water soluble ammonium nitrogen (NH_4^+) by titration (Mulvaney, 1996). Soluble and insoluble substances were determined by evaporation of the known amount of homogeneous sample filtrate on a water bath after absorption of insoluble substances on a filter, drying the evaporation residue at 105°C and determining its weight. COD was determined on the basis of organic substances oxidation in sample by potassium dichromate in sulfuric acid medium during 2-hour boiling in a COD reactor (HACH Company, Loveland, Colorado, USA). Portion of samples for N_t determinations were digested using a HACH-Digesdahl apparatus (HACH Company, Loveland, Colorado, USA). N_t was distilled with NaOH (40 %) (Bremner, 1996). The C content was calculated according to the content of OM by the method of Navarro et al. (1993) to obtain the C:N ratio.

2.3 Statistical analysis

The physical and chemical properties (pH, DM, IM, OM, NH_4^+, N_t,) of solid animal wastes, as well as the number of demaged eggs were expressed as mean values ± standard deviation ($\overline{x}$ ±SD).

Significance of differences between experimental and control groups of parasites were determined using Student t-test, ANOVA and Dunnet Multiple Comparison test at the levels of significance 0.05; 0.01 and 0.001 (Statistica 6.0).

Results

a) Anaerobic stabilisation of liquid animal wastes

Investigations were carried out under operating conditions of the large-capacity pig farm in Slovak Republic (Picture 3). Technological equipment for anaerobic treatment of pig slurry on the principle of methanogenesis with the production of biogas was built up on the farm (Picture 4). Pig slurry was treated in the bioreactor (2 500 m^3) manufactured by Mostáreň Brezno under the agreement with the firm BAUER Voitsberg. The stirring of the substrate in this reactor was done at the expense of energy of the generated biogas. Mean daily input of raw pig slurry in bioreactor of biogas plant varied between 78 and 144 m^3. The volume of digested slurry after methanogenesis was equal to that of the input. Two lagoons were the part of the biogas plant. The volume of larger lagoon is 20 000 m^3 (Picture 5) and that of smaller lagoon is 5 000 m^3. Both lagoons serve as reservoirs of digested slurry. Liquid fraction from the smaller lagoon was carried away and spread on fields. The presence and survival of parasite eggs were studied in the larger lagoon. Samples were taken from raw slurry collecting basin before the inlet in to bioreactor (input samples), from outlet of digested slurry after methanogenesis in bioreactor (output samples), from supernatant (liquid fraction) and from lagoon sludge (solid fraction - sediment). The slurry samples for parasitological and physical and chemical examination were collected monthly during 29 month.

Slurry from the pig farm stored in the collecting basin showed a considerable variability during the period of study (Table 1). Compared with mean pH value of 7.12 ± 0.26, pH raw slurry in the month 11, 17 and 21 was lower, ranging between 6.61 and 6.95. The most conspicuous differences were recorded in DM content, which is most likely associated with the amount of process water use. The DM content in raw slurry determined during the period studied ranged from 0.81 % to 5.30 %. The amount of NH_4^+ in raw slurry was between 821 mg.l^{-1} and 1 774 mg.l^{-1}. Chemical oxygen demand (COD) for that period varied from 2 000 mg.l^{-1} to 22 530 mg.l^{-1}. The mean contents of N_t, in slurry was 1 445 ± 420 mg.l^{-1}.

Picture 3. Large-capacity pig farm

Picture 4. Bioreactors of biogas plant

Picture 5. Large lagoon for storing digested pig slurry

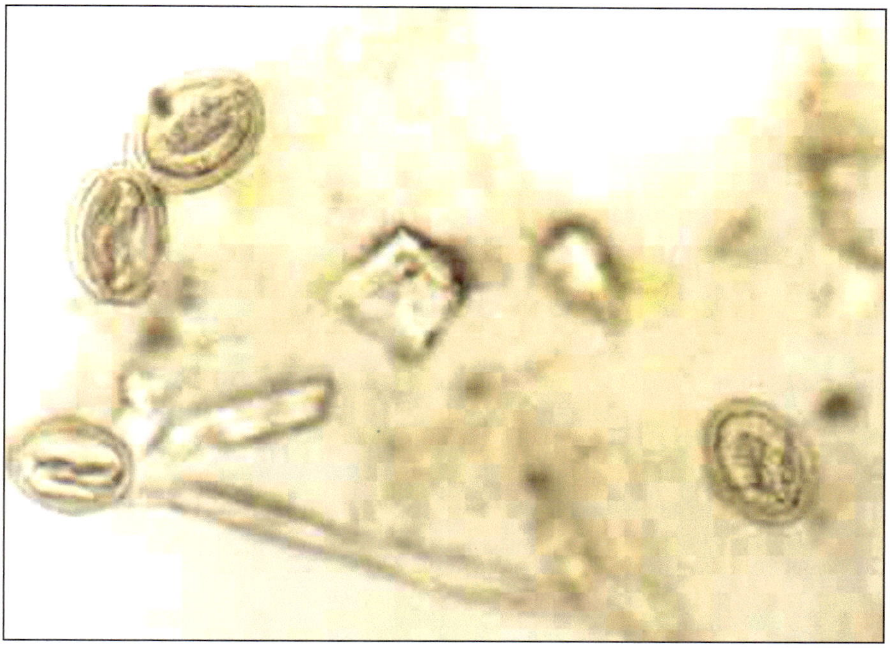

Picture 6. Embryonated *A. suum* eggs

Like raw pig slurry also slurry stabilised by anaerobic process showed variability of its physical-chemical parameters on its out flow from bioreactor (Table 2). Conspicuous differences were observed mainly in the dry mater content of anaerobically stabilized slurry.

This is caused by the projected input, reckoning on the 5 % of dry matter in raw pig slurry, but the mean dry matter content in raw slurry supplied to bioreactor was 1.96 % and therefor poultry excrements had to be regularly added (average DM content 22.27 %) to pig slurry prior to its supply into bioreactor. Stabilized slurry outlet of bioreactor contained as much as 3.23 ± 2.54 % DM on the average. Anaerobic digestion increased slurry pH which was ranging from 7.37 to 8.50. Compared with untreated slurry, anaerobic stabilization increased the content NH_4^+ to 7.80 ± 0.29 mg.l^{-1} on average. Concentration of N_t was increased twice.

Storage (month)	pH	COD (mg.l^{-1})	DM (%)	IM (%)	OM (%)	Soluble substances (mg.l^{-1})	Insoluble substances (mg.l^{-1})	NH_4^+ (mg.l^{-1})	N_t (mg.l^{-1})
0	7.44	14 833	2.75	31.87	68.13	11 263	16 264	1 774	2 419
1	7.34	2 000	0.84	51.42	48.58	5 836	2 612	1 186	1 401
2	7.17	9 297	0.95	43.04	56.96	4 561	4 897	821	1 195
3	7.03	13 500	1.14	57.71	42.29	7 757	3 641	1 202	1 485
4	7.00	20 900	1.57	38.74	61.26	11 095	4 572	1 078	1 363
5	7.35	14 824	0.81	45.71	54.29	4 895	3 178	1 037	1 191
6	7.36	13 333	2.52	17.36	52.64	5 366	19 844	1 247	1 429
11	6.61	21 795	5.30	33.02	66.98	-	-	1 695	1 089
17	6.95	12 750	0.95	30.53	69.47	1 000	8 500	1 478	1 010
21	6.95	22 530	2.80	19.97	80.03	5 870	22 130	1 358	1 872

Table 1. Physico-chemical properties of raw pig slurry (input sample of bioreactor) (COD – chemical oxygen demand, DM - dry matter, IM - inorganic mater, OM - organic matter; NH_4^+ - ammonium ions, N_t - total nitrogen, - - not examined)

Storage (month)	pH	COD (mg.l^{-1})	DM (%)	IM (%)	OM (%)	Soluble substances (mg.l^{-1})	Insoluble substances (mg.l^{-1})	NH_4^+ (mg.l^{-1})	N_t (mg.l^{-1})
0	8.50	36 333	-	-	-	-	-	2 633	6 320
1	7.74	10 500	0.81	56.54	43.46	4 739	3 401	2 204	2 605
2	7.63	17 820	1.24	48.50	51.50	6 134	6 226	2 157	2 699
3	7.80	8 500	1.96	59.69	40.31	6 192	13 456	2 045	2 549
4	7.69	17 100	3.16	41.81	58.19	5 965	5 658	1 933	3 138
5	7.77	6 092	4.48	42.06	57.94	3 225	41 603	1 898	1 982
6	7.92	2 186	2.91	42.87	57.13	3 555	25 518	2 437	3 516
11	7.88	4 872	0.50	70.00	30.00	-	-	2 171	1 530
17	7.37	7 750	6.45	39.84	60.16	1389	63 111	2 248	1 936
21	7.66	42 169	7.85	33.81	66.19	1 333	77 167	2 655	3 399

Table 2. Physico-chemical properties of digested pig slurry (output sample of bioreactor) (COD – chemical oxygen demand, DM - dry matter, IM - inorganic mater, OM - organic matter; NH_4^+ - ammonium ions, N_t - total nitrogen, - - not examined)

Anaerobically stabilized slurry was pumped from bioreactors into slurry ground lagoon for further storage. A long-term storage of digested slurry in lagoon is the most effective way of treatment resulting in a elimination of helminth eggs (Schwartzbrod et al., 1989). At the same time there is an increase in biogenic elements, especially of nitrogen and phophorus which are transformed into the forms acceptable by plants. Results of the chemical analysis of liquid fraction (supernatant) are presented in Table 3 and those of solid fraction (sludge) of lagoon in Table 4. pH of supernatant has not changed much over the period studied. Mean pH was 8.20 ± 0.11 %. Sediment pH decreased during the first period of the study (month 0-6) and than again increased. Ammonia content was about equal in both the fraction. The highest content of NH_4^+ was detected in spring month with its decrease observed in the course of study. N_t contained by supernatant samples varied between 882 mg.l^{-1} to 2 283 mg.l^{-1} (Table 3) and in sediment between 3 571 mg.l^{-1} to 57 831 mg.l^{-1} (Table 4). Sediment contained more DM and N_t than supernatant (Tables 3, 4).

Storage (month)	pH	COD (mg.l^{-1})	DM (%)	IM (%)	OM (%)	Soluble substances (mg.l^{-1})	Insoluble substances (mg.l^{-1})	NH_4^+ (mg.l^{-1})	N_t (mg.l^{-1})
0	8.30	4 500	0.50	61.06	38.94	4 416	581	1 737	1 910
1	8.20	4 000	0.50	70.39	29.61	4 808	174	1 307	1 428
2	8.17	2 002	0.68	57.58	42.42	6 579	239	1 345	1 569
3	8.34	3 500	0.66	57.75	42.25	5 340	1 272	1 111	1 214
4	8.10	7 600	0.93	56.04	43.96	6 085	3 177	1 408	1 662
5	8.08	6 552	0.87	52.76	47.24	3 600	5 255	1 135	1 172
6	8.29	1 530	0.71	57.59	42.41	2 748	4 337	1 107	1 223
13	8.21	7 059	0.70	55.53	44.47	5 954	1 083	1 863	2 283
14	8.07	818	1.68	46.60	53.40	5 588	11 217	1 569	1 569
15	8.28	1 904	0.66	54.83	45.17	5 325	1 284	1 331	1 317
16	8.21	5 385	0.63	56.38	43.62	4 483	1 806	896	882
17	8.29	8 605	0.60	54.27	45.73	3 501	2 524	616	1 415
23	8.32	3 333	0.35	71.43	28.57	2 128	1 372	672	1 016
29	7.95	5 000	0.75	45.33	54.67	3 000	4 500	862	1 031

Table 3. Physico-chemical properties of supernatant from stabilized pig slurry stored in lagoon (COD – chemical oxygen demand, DM - dry matter, IM - inorganic mater, OM - organic matter; NH_4^+ - ammonium ions, N_t - total nitrogen, - - not examined)

A. sum eggs and *Oesophagostomum* sp. eggs were rarely detected in slurry on the input and also on the output of bioreactor (Table 5). Similar results of helminths eggs occurrence in anaerobic slurry treatment were also presented by Juriš et al. (1996), No helminth eggs were found in the supernatant of digested slurry from the lagoon. *A. suum* eggs were found in sediment (Table 5).

High percentage of devitalised unembryonated *A. suum* eggs (47.46 ± 0.78 %) stored 11 months (from May – month 13 to March - month 23) in a ground slurry lagoon points to the impact of high concentration of NH_4^+ (max. 5 358 mg.l^{-1} in sediment compared to 1 863 mg.l^{-1} in supernatant), which are releasing during a period of time from an open area of the ground lagoon, and nitrogen (max. 9 854 mg.l-1 in sediment compared to 2 283 mg.l-1 in supernatant) on devitalization of developmental stages of endoparasites. The number of devitalised *A. suum* eggs increased towards to the bottom of lagoon. In the control groups, only 19.60 ± 1.80 % of *A. suum* eggs were devitalized (Table 6).

Storage (month)	pH	COD (mg.l-1)	DM (%)	IM (%)	OM (%)	Soluble substances (mg.l-1)	Insoluble substances (mg.l-1)	NH_4^+ (mg.l-1)	N_t (mg.l-1)
0	8.37	9333	1.17	49.21	50.79	1 885	2 138	5 778	5 963
1	8.13	11000	1.17	48.26	51.74	1 681	1 830	5 635	6 041
2	8.07	6170	1,70	43.21	56.79	1 643	2 241	7 344	9 652
3	8.09	4500	1.28	31.52	52.88	1 363	1 625	4 042	6 782
4	7.90	55100	1.12	34.90	65.10	1 359	2 437	3 913	7 298
5	8.08	8965	-	-	-	1 149	4 755	-	-
6	7.87	7322	-	-	-	1 541	-	-	-
13	-	-	-	-	-	-	--	-	-
14	7.73	6367	-	-	-	5 358	9 854	-	-
15	8.24	2494	1.72	32.11	67.89	840	1233	5513	11658
16	8.17	5897	0.73	51.16	48.84	1989	938	3740	3571
17	8.12	27186	1.19	45.41	54.59	915	1387	5402	6493
23	8.11	-	13.01	55.56	44.44	308	3909	72289	57831
29	-	-	-	-	-	-	-	-	-

Table 4. Physico-chemical properties of sediment from stabilized pig sllury stored in lagoon (COD – chemical oxygen demand, DM - dry matter, IM - inorganic mater, OM - organic matter; NH_4^+ - ammonium ions, N_t - total nitrogen, - - not examined)

Slurry	Storage (month) and occurence of eggs per litre sample															
	0	1	2	3	4	5	6	11	13	14	15	16	17	21	23	29
Input (raw)	Oe-2	ND	ND	A-5	ND	ND	ND	-	-	-	-	-	-	ND	ND	ND
Output (digested)	A-2	ND	ND	ND	ND	ND	ND	-	-	-	-	-	-	ND	A-1	ND
Supernatant (lagoon)	ND	ND	ND	ND	ND	ND	ND	ND	ND	ND	ND	ND	ND	ND	ND	ND
Sediment (lagoon)	ND	ND	A-6	ND	ND	ND	ND	ND	ND	ND	ND	ND	ND	A-2	ND	ND

Table 5. Occurence of helminth eggs in slurry and in lagoon (A – *A. suum* eggs, Oe – *Oesophagostomum* sp. eggs, ND – not detected, - - not examined)

Storage (month)	Damaged A. suum eggs ($\overline{x}$ %±SD)	
	Lagoon	Control
May (13)	16.23 ± 3.22	14.80 ± 2.43
June (14)	38.27 ± 2.51	15.79 ± 2.44
September (17)	40.37 ± 2.94	18.23 ± 1.22
March (23)	47.46 ± 0.78	19.60 ± 1.80

Table 6. Damage of *A. suum* eggs during long term storage of anaerobic stabilized pig slurry in lagoon

b) Anaerobic stabilisation of solid animal wastes

The effect of anaerobic stabilisation of solid animal wastes (manure, dog excrements) with or without addition of lime on the survival of parasitic germs were studied under laboratory conditions. Two types of lime was used in the experiment: 1. quality dust lime and 2. dust rejects from lime production caught on the electrostatic precipitator. General characteristics of tested lime are given in Table 7.

	Quality dust lime	Dust rejects
CaO + MgO	min. 95.0 %	min. 82.0 %
MgO	max. 5.0 %	max. 3.5 %
CO_2	max. 2.5 %	max. 11.0 %
Granularity	0-0.2 mm	0-1.0 mm

Table 7. Physico-chemical properties of the tested types of lime

Pig manure (M) and dog excrements mixed with hay in the ratio of 1:5 (D) were used in the experiment. Organic wastes were mixed with tested lime in a different concentration and periodically stirred. The following variations were investigated in comparison to untreated (control) manure (CM) and untreated dog droppings (CD):

a. manure mixed with quality dust lime in a concentration of 20 $g.kg^{-1}$ (ML20)
b. manure mixed with dust rejects in a concentration of 20 $g.kg^{-1}$ (M20)
c. dog droppings mixed with dust rejects in a concentration of 20 $g.kg^{-1}$ (D20),
d. dog droppings mixed with dust rejects in a concentration of 70 $g.kg^{-1}$ (D70).

Samples for parasitological and physical and chemical examinations were collected after 0, 1, 3, 8, 14, 36 (UM, ML20 and M20) and after 0, 1, 2, 3, 7, 8, 9, 10, 14, 73 (UD, D20, D70) days of exposure. Three samples were taken and analysed at each of the given sampling intervals.

The physical and chemical properties of treated manure and dog excrements are given in Tables 8 - 13. Comparison of the changes in The physical and chemical properties of organic material during anaerobic stabilisation with or withou dust rejects is given in Fig. 1 – 5.

Storage (days)	pH	DM (%)	IM (%)	OM (%)	NH_4^+ (mg.kg^{-1} DM)	N_t (mg.kg^{-1} DM)	C:N
0	8.47±0.58	33.22±6.88	8.09±2.57	91.91±2.57	120.89±7.05	13789.52±2356.62	34.27:1
1	8.57±0.02	20.66±4,29	10.41±0.72	89.58±0.72	257.65±10.10	51930.16±421.47	8.84:1
3	9.52±0.06	28.06±5.41	6.03±0.11	93.37±0.11	176.37±8.09	46522.24±2310.56	10.27:1
8	9.28±0.02	23.50±4.12	8.34±2.66	91.66±2.66	214.60±7.92	49872.94±1715.15	9.41:1
14	8.26±0.02	14.99±0.39	9.12±1.14	90.88±1.14	510.81±11.32	58608.01±2701.82	7.97:1
36	8.27±0.06	14.36±0.12	9.48±0.13	90.52±0.13	48.75±2.80	32698.26±2378.98	14.13:1

Table 8. Physico-chemical properties of the pig manure during anaerobic stabilization (CM) (DM - dry matter, IM - inorganic mater, OM - organic matter; NH_4^+ - ammonium ions, N_t - total nitrogen)

Storage (days)	pH	DM (%)	IM (%)	OM (%)	NH_4^+ (mg.kg^{-1} DM)	N_t (mg.kg^{-1} DM)	C:N
0	8.47±0.58	33.22±6.88	8.09±2.57	91.91±2.57	120.89±7.05	13789.52±2356,62	34.27:1
1	12.97±0.02	41.57±2.46	58.88±17.14	41.12±17.14	69.64±4.28	125901.56±873.31	8.11:1
3	12.76±0.01	45.96±3.72	39.91±7.76	60.09±7.76	111.75±9.84	18866,51±3349.86	16.32:1
8	10.39±0.01	26.91±2.16	16.95±2.67	83.05±2.67	236.01±7.05	52051.28±1482.36	8.17:1
14	8.29±0.01	21.47±5.22	14.36±3.41	85.64±3.41	326.22±17.94	56824.87±2746.13	7.72:1
36	8.29±0.01	20.11±2.32	13.41±1.12	86.19±1.12	225.21±22.47	71771.76±1722.51	6.14:1

Table 9. Physico-chemical properties of the pig manure mixed with dust rejects in a concentration of 20 g.kg^{-1} during anaerobic stabilization (M20) (DM - dry matter, IM - inorganic mater, OM - organic matter; NH_4^+ - ammonium ions, N_t - total nitrogen)

Storage (days)	pH	DM (%)	IM (%)	OM (%)	NH_4^+ (mg.kg^{-1} DM)	N_t (mg.kg^{-1} DM)	C:N
0	8.47±0.58	33.22±6.88	8.09±2.57	91.91±2.57	120.89±7.05	13789.52±2356,62	34.27:1
1	12.86±0.03	30.33±3.87	48.90±15.23	51.10±15.23	126.24±9.84	37815.69±1860.53	6.97:1
3	12.96±0.01	37.31±3.89	57.40±5.92	42.60±5.92	130.15±9.01	35790.24±2332.63	6.08:1
8	11.56±0.02	25.37±0.95	48.67±3.35	51.33±3.35	176.67±10.10	81616.08±3704.40	3.21:1
14	9.36±0.01	20.30±2.17	34.12±1.12	65.88±1.12	206.99±17.83	44057.78±2515.94	7.66:1
36	8.76±0.01	20.08±1.56	32.48±3.46	67.52±3.46	181.37±25.75	65746.86±2677.51	5.25:1

Table 10. Physico-chemical properties of the pig manure mixed with quick lime in a concentration of 20 g.kg^{-1} during anaerobic stabilization (ML20) (DM - dry matter, IM - inorganic mater, OM - organic matter; NH_4^+ - ammonium ions, N_t - total nitrogen, - - not examined)

Storage (days)	pH	DM (%)	IM (%)	OM (%)	NH_4^+ (mg.kg^{-1} DM)	N_t (mg.kg^{-1} DM)	C:N
0	9.08±0.01	35.66±1.83	11.33±0.01	88.67±0.01	219.07±55,70	40758.43±1416.02	11.15:1
1	8.57±0.01	34.66±0.11	14.29±1.77	85.71±1.77	232.05±23,57	39116.17 ±207.87	11.24:1
2	9.61±0.01	35.20±4.23	19.21±4.89	80.79±4.89	395,72±2,48	41116.07±1205.26	10.12:1
3	9.78±0.01	37.56±1.93	22.33±1.06	77.67±1.06	309.78±95.04	44207.73±3222.05	9.05:1
7	9.01±0.01	37.17±0.29	20.58±0.48	79.42±0.48	370,89±8.22	23346.91±5147.86	18.06:1
8	9.39±0.02	33.21±0.18	22.46±1.97	77.54±1.97	82.25±2.55	12152.63±77.34	32.74:1
9	9.55±0.02	29.81±3.03	31.17±3.54	68.83±3.54	132.58±72.91	6556.92±818.39	54.92:1
10	9.49±0.03	31.96±1.13	31.46±0.36	68.54±0,36	124.07±0.16	6217.49±27.34	57.31:1
14	9.34±0.03	53.90±4.33	32.32±0.91	67.68±0.91	138.03±7.41	5435.17±2904.64	76.40:1
73	8.51±0.03	86.31±0.23	17.99±1.73	82.01±1.73	28.09±3.26	9159.80±1327.01	46.33:1

Table 11. Physico-chemical properties of the dog excrements during anaerobic stabilization (CD) (DM - dry matter, IM - inorganic mater, OM - organic matter; NH_4^+ - ammonium ions, N_t - total nitrogen, - - not examined)

Storage (days)	pH	DM (%)	IM (%)	OM (%)	NH_4^+ (mg.kg^{-1} DM)	N_t (mg.kg^{-1} DM)	C:N
0	8.41±0.05	37.21±0.01	16.04±4.81	83.96±4.81	400.57±47.84	45177.35±4724.10	9.64:1
1	11.21±0.02	44.47±0.91	39.37±0.10	60.63±0.10	12.51±8.65	32262.40±4212.96	9.66:1
2	9.34±0.03	56.45±15.15	62.63±19.97	37.37±19.97	36.17±7.84	16218.95±2664.67	12.44:1
3	8.58±0.01	57.28±33.23	60.29±28.06	39.71±28.06	645.10±362.56	45266.01±19058.77	4.23:1
7	9.08±0.01	45.11±6.79	43.58±0.64	56.42±0.64	225.45±91.01	22104.48±11603.90	15.37:1
8	9.13±0.01	43.90±2.84	41.87±0.96	58.13±0.96	439.78±141.28	18254.61±1760.78	16.74:1
9	9.27±0.05	68.66±1.32	27.61±0.60	72.39±0.60	398.59±2.51	18083.74±303.53	20.86:1
10	9.12±0.04	64.15±0.16	42.28±0.71	57.64±0.71	349.02±10.08	16966.79±191.73	17.70:1
14	8.91±0.03	60.02±0.98	49.20±4.49	50.80±4.49	338.50±24,95	17963.19±457.92	14.71:1
73	8.69±0.01	89.06±0.01	38.85±3.90	61.15±3.90	74.54±26.60	13972.75±1214.80	22.56:1

Table 12. Physico-chemical properties of the dog excrements mixed with dust rejects in a concentration of 20 g.kg^{-1} during anaerobic stabilization (D20) (DM - dry matter, IM - inorganic mater, OM - organic matter; NH_4^+ - ammonium ions, N_t - total nitrogen, - - not examined)

Storage (days)	pH	DM (%)	IM (%)	OM (%)	NH_4^+ (mg.kg^{-1} DM)	N_t (mg.kg^{-1} DM)	C:N
0	9.08±0.01	35.66±1.83	11.33±0.01	88.67±0,01	219.07±55.70	40758.43±1416.02	11.15:1
1	12.58±0.04	43.11±1.51	58.48±9.02	41.52±9,02	41.04±42.79	13041.67±2498.12	16.22:1
2	12.68±0.01	46.81±0.17	62.89±2.38	37.11±2,38	10.19	4555.21±936.46	42.69:1
3	12.64±0.01	44.10±1.95	57.53±1.39	42.47±1,39	20.65±14.40	6868.28±1649.01	32.67:1
7	12.36±0.01	45.50±0.21	60.84±0.47	39.16±0,47	140.34±9.77	26830.72±6801.00	7.73:1
8	10.63±0.01	45.22±0.48	62.63±2.26	37.37±2,26	131.32±2.98	14216.80±5981.61	15.04:1
9	10.12±0.01	45.60±1.23	60.40±4.67	39.60±4.67	82.59	16161.23±10202.44	16.18:1
10	10.06±0.02	46.75±1.09	63.09±0.21	36.91±0.21	85.65±10.62	13399.86±759.04	14.14:1
14	9.82±0.01	52.21±1.08	65.03±0.80	34.97±0.80	14.28	19371.98±1147.30	9.21:1
73	8.97±0.02	87.17±0.47	40.50±1.92	59.50±1.92	2.20±0.08	12440.27±566.93	24.43:1

Table 13. Physico-chemical properties of the dog excrements mixed with dust rejects in a concentration of 20 g.kg^{-1} during anaerobic stabilization (D70) (DM - dry matter, IM - inorganic mater, OM - organic matter; NH_4^+ - ammonium ions, N_t - total nitrogen, - - not examined)

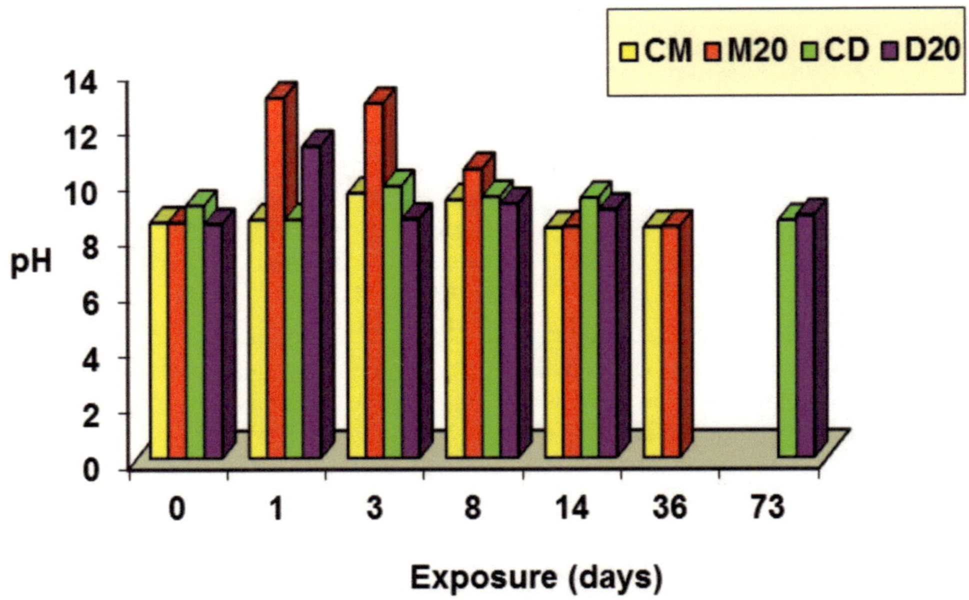

Fig. 1. Comparison of the changes in pH of organic material during anaerobic stabilisation with or without dust rejects

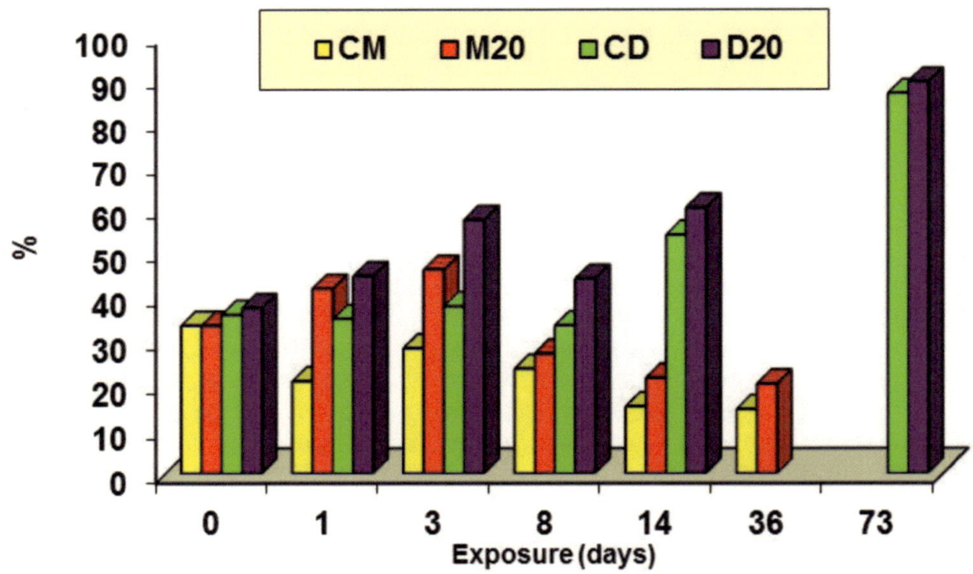

Fig. 2. Comparison of the changes in DM of organic material during anaerobic stabilisation with or without dust rejects

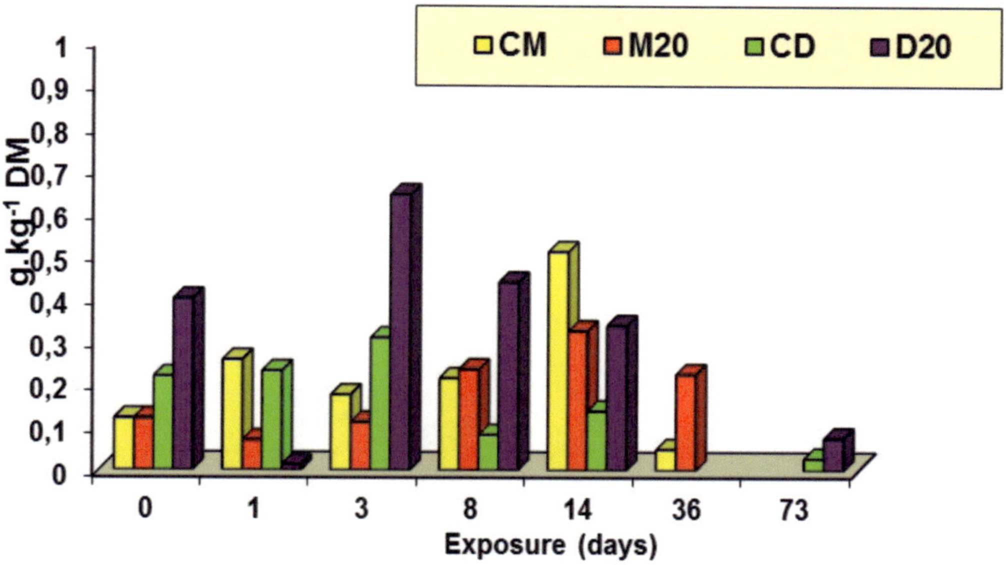

Fig. 3. Comparison of the changes in NH_4^+ of organic material during anaerobic stabilisation with or without dust rejects

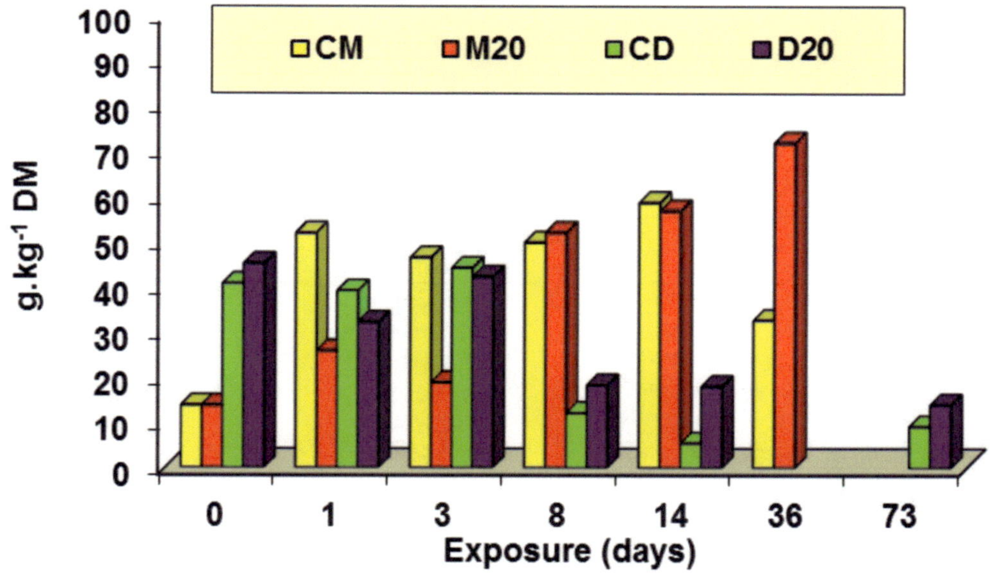

Fig. 4. Comparison of the changes in N_t of organic material during anaerobic stabilisation with or without dust rejects

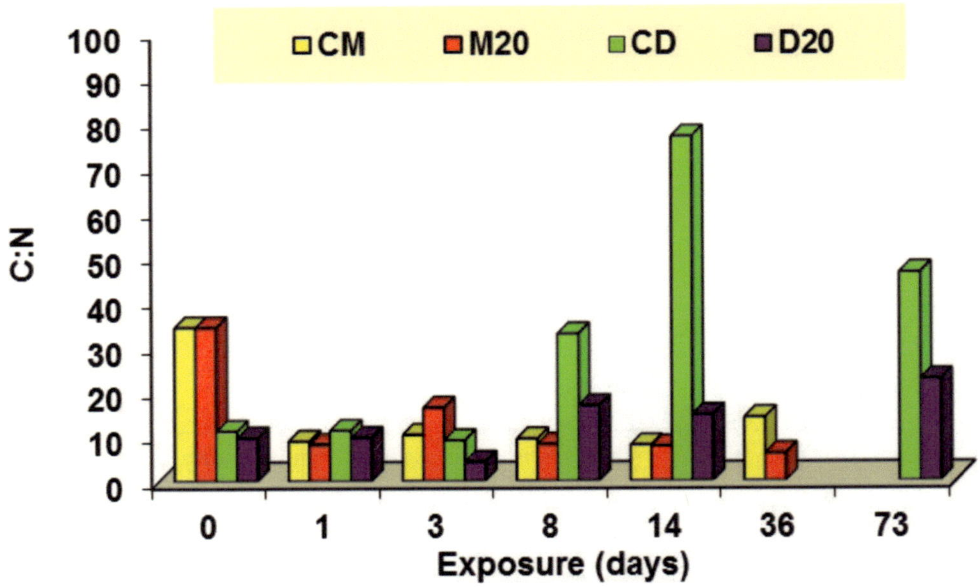

Fig. 5. Comparison of the changes in C:N ratio of organic material during anaerobic stabilisation with or without dust rejects

The Table 14 indicates that a 24 hour after application of both types of lime at concentration 20 g.kg^{-1} manure more than 80 % of model unembryonated *A. suum* eggs were devitalised. *A. suum* eggs were totally devitalised as early as till 36 days after application of lime in manure. 58.13±6.89 % of eggs were devitalised in the control without dust reject in the end of experiment.

Storage (days))	Demaged *A. suum* eggs (x%±SD)		
	CM	ML20	M20
0	16.43±1.14	16.43±1.14	16.43±1.14
1	36.31±2.46	82.41±8.49***	80.68±6.75***
3	55.10±10.72**	87.23±11.06***	89.85±5.10***
8	59.14±1.74**	98.96±1.80***	82.22±16.78***
14	56.11±19.64*	97.13±3.77***	97.33±4.62***
36	58.13±6.89**	100***	100***

Table 14. Survival of *A. suum* eggs during anaerobic stabilisation of the dog excrements with or without lime (* Significance at the level P<0.05; ** Significance at the level P<0.01; *** Significance at the level P<0.001)

For the sanitation of animal excrenemts, the use of dust rejects from lime production, at more affordable price than quality lime, is very suitable. An application of dust rejects to the mixed dogs' excrements at a concentration of 20 g.kg^{-1} of organic wastes, resulted in a devitalisation of 65.65 ± 2.84 % and at a concentration of 70 g.kg^{-1} 77.05 ± 2.36 % of model unembryonated *A. suum* eggs within 24 hours (Table 15). *A. suum* eggs were totally devitalised as early as within 8 days in dogs' excrements after application of dust rejects at a concentration of 70 g.kg^{-1} and within 21 days after application of dust rejects at a concentration of 20 g.kg^{-1} due to the changes in physical and chemical properties of the stabilised materials (Tables 12, 13). 57.23±3.21 % of eggs were devitalised in the control without dust reject in the end of experiment (Table 15).

Storage (days)	**Demaged *A. suum* eggs (x%±SD)**		
	CD	P20	P70
0	12.62±1.14	12.62±1,14	12.62±1.14
1	35.70±2.46	65.65±2.84**	77.05±2.36***
2	54.43±10.66*	68.65±3.89**	82.30±4.81***
3	67.00±2.55**	75.15±1.21**	87.60±3.98***
7	62.65±4.03**	76.25±5.41**	97.13±3.97***
8	59.80±2.71*	76.93±2.69***	100***
9	61.87±2.90*	82.30±4.81***	100***
10	62.85±4.03*	85.69±1.45***	100***
14	61.96±3.26*	95.69±6.35***	100***
21	55.65±2.36*	100***	100***
73	57.23±3.21*	100***	100***

Tab. 15. Survival of *A. suum* eggs during anaerobic stabilisation of manure with or without lime (* Significance at the level $P<0.05$; ** Significance at the level $P<0.01$; *** Significance at the level $P<0.001$)

Our experiment showed that stabilisation of organic wastes with dust rejects result in complete devitalisation of *A. suum* eggs (Table 14, 15). The most important physico-chemical factors affecting viability of helminth eggs include pH and ammonia. We observed the highest pH and ammonia content especially in the organic wastes treated with tested types of lime. One of our previous studies (Ondrašovič et al., 2002) on the effect of ammonium hydroxide on *A. suum* eggs showed that at 10 % concentration of NH_4OH, pH 12.16 and exposure time 180 min. approx. 94 % *A. suum* eggs were devitalised. Pescon and Nelson (2005) also reported that environmentally relevant concentrations of ammonia may significantly increase the rate of *Ascaris* eggs inactivation during alkaline stabilization.

3. Conclusion

Processes of slurry anaerobic stabilization represent an effective method in terms of energy, since the substantial portion of energy present in easily decomposable organic constituents of the substrate is acquired in the form of biogas. Non-decomposed organic matter is well stabilized from hygienic point of view. Anaerobic stabilization increases the proportion of biogenic element (especially nitrogen) converting stabilized excrements into quality fertilizer. Anaerobically stabilized pig slurry stored in lagoon significantly influence the quality and quantity of grasses, depending on the dose of slurry used and on weather conditions. From the nutritional point of view, the sludge (sediment) from ground lagoon is also important for plants (Valocká et al., 2000). The high amount of nitrogen is apparently the result of the decomposition process going on in lagoon.

Hazard of contamination of field fertilized with the lagoon effluent increases when raw slurry is used for fertilization of soil or pastures. When slurry is processed in a wastewater treatment plant, parasitic eggs concentrated in solid fraction. It is therefore necessary to pay a proper attention to slurry processing.

The anaerobic stabilization and the use of dust rejects from lime production, at more affordable price than quality lime dust, were demonstrated to be very suitable for the sanitation of organic wastes from animal production and dog excrements. This way of treatment is thus not associated with a risk of dissemination, survival and potential spread of developmental stages of endoparasites to the environment via stabilized organic wastes

4. Acknowledgements

We would like to thank CARMEUS Ltd. for providing lime and people who allowed us to gather data needed for elaboration of this chapter. This study was financially supported by the project VEGA No. 2/0147/10.

5. References

Antolová, D., Reiterová, K., Miterpáková, M., Stanko, M., Dubinský, P. (2004). Circulation of *Toxocara* spp. in suburban and rural ecosystems in the Slovak Republic. In *Veterinary Parasitology*. Vol. 126, no. 3, p. 317-324.

Bremner, J. M. (1996). Nitrogen - total. In D. L. Sparks (Ed.), *Methods of Soil Analysis* (pp. 1085-1121). Madison, WI: SSSA Inc..

Crompton, D. W. (1999). How much human helminthiasis is there in the world? *Journal of Parasitology, 85*, 397-403.

Cherepanov, A. A. (1982). Methods of Laboratory Centrals of Cleaning Plants on Farms. Kolos. Moscow (In Russian).

Despommier, D. (2003). Toxocariasis: Clinical aspects, epidemiology, medical ecology and molecular aspects. Clin. Microbiol. Rev. 16, 265-272.

Dubinský, P., Krupicer, I., Levkut, M., Švický, E., Dvorožňáková, E., Revajová, V., Vasilková, Z., Kováč, G., Reiterová, K., Lenhardt, Ľ., Ondrejka, R., Papajová, I., Moncol, D. J. (2000). Influence of *Ascaris suum* infection on ruminants. In: Dubinský, P., Juriš, P., Moncol, D. J. (Eds.): Environmental protection against the spread of pathogenic agents of diseases through the wastes of animal production in the Slovak Republic. Harlequine, Ltd., Košice, p. 143-160.

Eckert, J. (1992). Dauerformen von Parasiten als umwelthygienisches Problem, in: Eckert, J., Kutzer, E., Rommel, M., Bürger, H.J., Körting, W. (Eds.), Veterinärmedizinische Parasitologie, 4. Aufl., Verlag Paul Parey, Berlin, 87-107.

Juriš, P., Vasilková, Z., Krupicer, I., Plachý, P., Sasáková, N. (1996). Hygienic and ecological aspects of mesophilic stabilization of swine liquid excrements under the operating conditions. In Hygienic and ecological problems in relation to veterinary Medicine, Košice, Slovak republic, p. 119-128 (In Slovak).

Juriš, P., Rataj, D., Ondrašovič, M., Sokol, J., Novák, P. (2000). Sanitary and ecological requirements on recycling of organic wastes in agriculture. Vyd. Michala Vaška, Prešov, 1-178 (in Slovak).

Krupicer, I., Valocká, B., Vasilková, Z., Sabová, M., Papajová, I., Dubinský, P. (2000). Contamination and survival of helminth eggs in pig slurry and influence of the lagoon effluent on soil and plant parasitic nematodes. In: Dubinský, P., Juriš, P., Moncol, D. J. (Eds.): Environmental protection against the spread of pathogenic agents of diseases through the wastes of animal production in the Slovak Republic. Harlequine, Ltd., Košice, p. 79-93.

Lauková, A., Juriš, P., Vasilková, Z., Papajová, I. (2000). Treatment of sanitary-important bacteria by bacteriocin substance V24 in cattle dung water. Letters in Applied Microbiology, 30, p. 402-405.

Matsuo, J., Nakashio, S. (2005). Prevalence of fecal contamination in sandpits in public parks in Sapporo City, Japan. In *Veterinary Parasitology*, Vol. 128, p. 115-119.

Miterpáková, M., Dubinský, P. Reiterová, K., Stanko, M. (2006) Climate and environmental factors influencing *Echinococcus multilocularis* occurrence in the Slovak Republic. In *Annals of Agricultural and Environmental Medicine*. Vol. 13, no. 1, p. 235-242.

Mulvaney, R. L. (1996). Nitrogen - inorganic forms. In D. L. Sparks (Ed.), *Methods of Soil Analysis* (pp. 1123-1184). Madison, WI: SSSA Inc..

Navarro, A. F., Cegarra, J., Roig, A., Garcia, D. (1993). Relationships between organic matter and carbon contents of organic wastes. *Bioresource Technology, 44*, 203-207.

Ondrašovič, M., Juriš, P., Papajová, I., Ondrašovičová, O., Ďurečko, R., Vargová, M. (2002): Lethal effect of selected disinfectants on *Ascaris suum* eggs. Helminthologia, 39, pp. 205-209.

Papajová, I., Juriš, P. (2009). The effect of composting on the survival of parasitic germs. In: Pereira, J. C., Bolin, J. L. (Eds.) Composting: Processing, Materials and Approaches. New York : Nova Science Publishers, p. 124-171. ISBN 978-1-60741-438-4.

Pescon, B. M., Nelson, K. L. (2005). Inactivation of *Ascaris suum* eggs by ammonia. Environ. Sci. Technol., 39, pp. 7909-7914.

Plachý, P., Juriš, P. (1995). Use of polyurethane carrier for assessing the survival of helminth eggs in liquid biological sludges. Vet. Med. 40, 323-326.

Sasáková, N., Juriš, P., Papajová, I., Vargová, M., Ondrašovičová, O., Ondrašovič, M., Kašková, A., Szabová, E. (2005). Parasitological and bacteriological risks to animal and human health arising from waste-water treatment plant. Helminthologia, 42, p. 137-142.

Schwartzbrod, J., Stien, J. L., Bouhoum, K., Baleux, B. (1989). Impact of wastewater treatment on helminth eggs. Water Science and Technology, 21, 295-297. STATISTICA 6.0, StatSoft Inc., USA.

STATISTICA 6.0, StatSoft Inc., USA.

Tofant, A., Vučemilo, M., Hadžiosmanović, M., Križanić, J. (1999). Liquid manure: A surface water pollutant [Einfluß der düngung landwirtschaftlicher flächen mit schweinegülle auf die wasserqualität in naheliegenden gewässern] *Tierarztliche Umschau* 54, 148-150.

Valocká, B., Dubinský, P., Papajová, I., Sabová, M. (2000). Effect of anaerobically digested pig slurry from lagoon on soil and plant nematode communities in experimental conditions. Helminthologia, 37, p. 53-57.

4

The Waste Oil Resulting from Crude Oil Microbial Biodegradation in Soil

Anatoly M. Zyakun, Vladimir V. Kochetkov
and Alexander M. Boronin
Skryabin Institute of Biochemistry and Physiology of Microorganisms RAS
Russia

1. Introduction

Environmental pollution by oil and oil products, which occurs at petroleum extraction wells, as a result of spills from oil tankers, pipe line breaks, disposal of refinery waste, leaks at gasoline stations, etc., have caused tremendous damage to ecological systems especially to many plant species (Adam and Duncan 2002; 2003; Palmroth et al. 2005), and a wide array of animals (Khan and Ryan 1991; Tevvors and Sair 2010). According to available data (Wang et al. 2011), the total amount of all major spills in the world was about 37 billion barrels of crude oil pollute soil and water ecosystems. It exceeds the total amount of crude oil consumption for the entire world annually (30 billion barrels in 2006) (Mundi 2010). Consequently, the problem of environmental pollution with anthropogenic hydrocarbons and their influence on natural ecosystems calls for comprehensive investigation. Crude oil consists of a number of rather complicated components, which are toxic and can exert side effects on environmental systems. Oil pool contains aliphatic and polycyclic aromatic hydrocarbons, for example, crude oil consists of alkanes 15 - 60 %, naphthenes 30-60 %, aromatics 3-30% and asphaltenes 6 % by weight (Speight 1990). The extent of oil spills can have a legacy for decades, evens centuries in future (Wang et al. 2011). Toxic effects of oil and oil products on the soil environment include increasing hydrophobicity of soils and disruption of water availability to vegetation, and direct toxicity to plants and microorganisms. At the sub-toxic level, negative effects may include the absorption of low-molecular oil hydrocarbons into plant tissues, and the inhibition or activation of microbial soil processes. The soil, although is an important sink for a wide range of substances, pollutant load exceeding certain threshold has the potential of impacting negatively on the capacity of the soil to perform its ecosystem functions with repercussions on sustainability issues such as plant growth and some non-hydrocarbon utilizing microorganisms. For instance, the aromatics in crude oil produce particular adverse effect to the local soil microbiota. It was found that phenolic and quinonic naphthalene derivatives inhibited the growth of some microbial cells (Sikkema et al. 1995). As follows from the work (Wongsa et al. 2004), the rates of utilization of separate oil fractions may be significantly differed even in case of one and the same strain of hydrocarbon-oxidizing microorganisms. As a result, the influence of microorganisms on crude oil in soil may be accompanied by substantial changes in the initial composition of hydrocarbons, while the rest of hydrocarbons in soil may have absolutely different properties compared to the initial characteristics. The term 'waste oil' was used to designate the hydrocarbon tails of crude oil introduced into soil and transformed into the product that lost the original properties (i.e., the quantitative ratio of hydrocarbon components changed and the organic products of microbial biosynthesis appeared, which differ from the initial oil components in metabolic availability for a wide range of soil microorganisms, etc). It has been known that soil microbial communities are able to adjust to unfavourable conditions and to use a broad spectrum of substrates (Jobson et al. 1974; Nikitina et al. 2003). They have unique metabolic systems that allow them to utilise both natural and anthropogenic substances as a source of energy and tissue constituents. These unique characteristics make the microbiota useful tool in monitoring and remediation processes. Bioremediation of soil contaminated with oil hydrocarbons has been established as an efficient, economic, versatile, and environmentally sound treatment (van Hamme et al. 2003). Several reports have already focused on the composition of natural microbial populations contributing to biotransformation and biodegradation processes in different environments polluted with hydrocarbons (Juck et al. 2000; Hamamura et al. 2008; Marques et al. 2008). It is becoming increasingly evident that the fate of anthropogenic hydrocarbons pollutants entering the soil system requires efficient monitoring and control. The bioremediation potential of microbial communities in soil polluted with oil hydrocarbons depends on their ability to adapt to new environmental conditions (Mishra et al. 2001; Kaplan and Kitts 2004). Investigations into how bioremediation influences the response of a soil microbial community, in terms of activity and diversity, are presented in a series of publications (Jobson et al. 1974; Margesin and Schinner 2001; Zucchi et al. 2003; Hamamura et al. 2006; Margesin et al. 2007). The methods of monitoring and characterization of hydrocarbon degrading activity of soil microbiota are of special interest (Margesin and Schinner 2005; Abbassi and Shquirat 2008; Pleshakova et al. 2008). Oil hydrocarbon biodegradation and transformation in soils can be monitored by estimating the concentration of pollutant (Tzing et al. 2003) and the formation of respective metabolites. The most ubiquitous and universal metabolites is carbon dioxide (CO_2), since respiration is by far the prominent pathway of biologically processed carbon.

The activity of soil microbiota can be characterized by the method of the substrate-induced respiration (SIR) which was used for the measurement of CO_2 production and the estimation of soil microbial biomass. When an easily microbial degradable substrate, such as glucose, is added to a soil, an immediate increase of the respiration rate is obtained, the size of which is assumed to be proportional to size of the microbial biomass (Anderson and Domsch 1978). In addition to SIR, the index of the specific microbial activity in soil is the priming effect (PE) of introduced exogenous substrate, which was defined as 'the extra decomposition of native soil organic matter in a soil receiving an organic amendment" (Bingeman et al. 1953). The PE may be represented by the following three indices: (a) positive PE shows that exogenous substrate introduction concurrent with its mineralization increases SOM mineralization to a rate exceeding the previous rate; (b) zero PE shows that CO_2 is produced additionally only as a result of microbial mineralization of introduced substrate without changing the existing rate of SOM mineralization; and (c) negative PE values show that exogenous substrate introduction decreases SOM mineralization rate and CO_2 production is determined mainly by mineralization of the substrate. PE determination only by the difference of CO_2 production rate before and after substrate introduction into soil suffers from the known uncertainly of CO_2 sources and does not allow distinguishing between the so-called "real" and "apparent" PE. (Blagodatskaya et al. 2007; Blagodatskaya and Kuzyakov 2008). Obviously, unambiguous determination of PE by CO_2 production calls for an exogenous substrate different from SOM in carbon isotopes (Zyakun et al. 2003; Dilly and Zyakun 2008; Zyakun et al. 2011). It has been shown that addition to the soil of a substrate easily accessible for microorganisms (e.g., glucose, amino acids, etc.) (Harabi and Bartha1993; Shen and Bartha 1996; Zyakun and Dilly 2005; Blagodatskaya and Kuzyakov 2008), contributes to the increase of SOM mineralization rate 2-3-fold compared to the processes in native soil. Acceleration of SOM degradation (positive PE) was also observed in case of addition of an aliphatic hydrocarbon (n-hexadecane) to the soil. Introduction into soil of n-hexadecanoic acid, the product of n-hexadecane oxidation, resulted in the lower rate of SOM mineralization compared to native soil (negative PE) (Zyakun et al. 2011). In the light of brief presentation of methods characterizing biodegradation and transformation of exogenous organic products entering the soil, the fate of crude oil in soils may be defined by the following parameters: (a) the rate of CO_2 production as result of mineralization of crude oil and SOM; (b) activation of mineralization of native soil organic matter by introduced substrate (priming effect); c) the ratio of the quantities of biomass of the microorganisms growing on oil hydrocarbons as a substrate and quantities of SOM mineralized into CO_2.

2. Methods used to analyze the CO_2 microbial production in soil

2.1 CO_2 sampling

Soil samples, 100 g dry weight, were placed into 700-ml glass vials, hermetically closed and pre-incubated for 3 days at 22 ^{0}C. Metabolic carbon dioxide (CO_2) formed by microbial mineralization of SOM and test-substrate (crude oil) was collected using glass plates (10 ml) placed the over soil surface, containing 2-3 ml of 1M NaOH solution. Production of CO_2 in the course of the experiment in each of the vials was determined by titration of the residual alkali in the plates using an aqueous 0.1M HCl solution. The total amount of CO_2 fixed in the NaOH solution was also determined by precipitation with $BaCl_2$ and quantitative retrieval of $BaCO_3$. Barium carbonate was washed with water, precipitated, dried, and the resulting precipitate weighed and used for quantitative calculation of metabolic CO_2 production and carbon isotope analysis.

2.2 The kinetics of CO_2 respiration

Specific CO_2 evolution rates (μ) of soil microorganisms after crude oil addition to soil were estimated from the kinetic analysis of substrate-induced respiration ($CO_2(t)$) by fitting the parameters of equation [1]:

$$CO_2(t)=K+r \cdot exp(\mu \cdot t) \tag{1}$$

where K is the initial respiration rate uncoupled from ATP production, r is the initial rate of respiration by the growing fraction of the soil microbiota which total respiration coupled with ATP generation and cell growth, and t is time (Panikov and Sizova 1996; Stenström et al. 1998; Blagodatsky et al. 2000). The lag period duration (t_{lag}) was determined as the time interval between substrate addition and the moment when the increasing rate of microbial growth-related respiration $r \cdot exp(\mu \cdot t)$ became as high as the rate of respiration uncoupled from ATP generation.

$$t_{lag}=\ln(K/r)/\mu \quad (2)$$

According to the theory of microbial growth kinetic (Panikov 1995; Blagodatskaya et al. 2009), the lag period was calculated by using the parameters of approximated curve of respiration rate of microorganisms with [2].

2.3 Carbon isotopic analysis

The metabolic activity of soil microbial community with respect to substrate (crude oil hydrocarbons) was determined from CO_2 evolution rates and the ^{13}C-CO_2 isotope signature. The characteristics of abundance ratios of carbon isotopes $^{13}C/^{12}C$ in SOM, crude oil, and metabolic CO_2 (as $BaCO_3$) were measured using by isotopic mass-spectrometry (Breath MAT-Thermo Finnigan) connected with a gas chromatograph via ConFlow interface. Isotope analysis of metabolic CO_2 was performed using about 3-4 mg of obtained $BaCO_3$ [M = 197.34], which then was degraded to CO_2 by orthophosphoric acid in a 10-ml container. For the analysis of carbon isotope contents of organic matter, SOM and crude oil samples were combusted to CO_2 in ampoules at 560 ^{0}C in the presence of copper oxide.

The ratios of peak intensities in CO_2 mass spectra with m/z 45 ($^{13}C^{16}O_2$) and 44 ($^{12}C^{16}O_2$) were used for quantitative characterization of the content of ^{13}C and ^{12}C isotopes in the analyzed samples. According to the accepted expression [3], the amount of ^{13}C isotope was determined in relative units $\delta^{13}C$ (‰):

$$\delta^{13}C = (R_{sa}/R_{st} - 1)\ 1000\ ‰ \quad (3)$$

where $R_{sa}=(^{13}C)/(^{12}C)$ represented the abundance ratios of isotopes $^{13}C/^{12}C$ in a sample and $R_{st}=(^{13}C)/(^{12}C)$ was the ratio of these isotopes in the International Standard PDB (Pee Dee Belemnite) (Craig 1957). Each CO_2 sample was analyzed in three repeats; standard error was about ± 0.1‰. The $\delta^{13}C$ values are characteristics of stable isotope composition or the $^{13}C/^{12}C$ abundance ratio in the analyzed compounds. Negative values indicate the ^{13}C depletion; positive values indicate ^{13}C enrichment relative to PDB standard.

2.4 Mass isotope balance

Metabolic carbon dioxide produced in the experiments and controls was accumulated during the appropriate time intervals (1-3 days) followed by determination of its quantity and carbon isotope characteristics. The average weighed carbon isotope composition of metabolic CO_2 ($\delta^{13}C_{ave}$), which was obtained in detached time intervals, was determined using the expression [4]:

$$\delta^{13}C_{ave}= (\sum q_i \cdot \delta^{13}C_i)/\sum q_i,\ ‰ \quad (4)$$

where q_i and $\delta^{13}C_i$ were CO_2 production rate and carbon isotope composition at *i*–intervals, respectively.

Determination of mass isotope balance is based on the suggestion that the characteristics of carbon isotope content ($\delta^{13}C$) of CO_2 produced during microbial mineralization of hydrocarbons will inherit the $\delta^{13}C$ value of crude oil with an accuracy of isotopic fractionation effect. According to (Zyakun et al. 2003), the $\delta^{13}C$ value of metabolic CO_2

produced during oxidation of n-hexadecane and aliphatic hydrocarbons was less by 1-3 ‰ compared to the isotope characteristics of substrates used. It means that the $\delta^{13}C$ value of CO_2 produced during microbial degradation of oil hydrocarbons was estimated by $\delta^{13}C$ equal to the value over a rang of -28 to -31 ‰, where $\delta^{13}C$ of the crude oil was about of $\delta^{13}C_{oil}$ = –28,4±0,2 %o. It is rather different from CO_2 resulting from soil organic matter (SOM) mineralization ($\delta^{13}C_{SOM}$ is equal to -23,5±0,5 ‰ for the soil). Thus, after addition of the oil hydrocarbon to soil, the mass isotope balance for CO_2 evolved during microbial mineralization of SOM and the exogenous substrate (SUB) was calculated using equation [5]:

$$\delta^{13}C_{SOM} \times Q_{SOM} + \delta^{13}C_{SUB} \times Q_{SUB} = \delta^{13}C_{MIX} \times (Q_{SOM} + Q_{SUB}) \quad (5)$$

where $\delta^{13}C_{SOM}$ and $\delta^{13}C_{MIX}$ are isotopic characteristics of ^{13}C content in CO_2 before and after substrate addition to the soil; $\delta^{13}C_{SUB}$ is the isotopic characteristic of ^{13}C content in CO_2 produced during microbial mineralization of the test substrate; Q_{SOM} and Q_{SUB} are CO_2 quantities resulted from microbial mineralization of SOM and added substrate in the soil samples, respectively.

Here the shares of CO_2 formed by mineralization of SOM (F_{SOM}) and oil hydrocarbons (F_{SUB}) are presented, by definition, as [6] and [7]:

$$F_{SOM} = Q_{SOM} / (Q_{SOM} + Q_{SUB}) \quad (6)$$

$$F_{SUB} = (1 - F_{SOM}) = Q_{SUB} / (Q_{SOM} + Q_{SUB}) \quad (7)$$

Using carbon isotope characteristics of total CO_2 formed by microbial mineralization of SOM and oil hydrocarbons ($\delta^{13}C_{tot}$) (in experiments) and CO_2 formed by mineralization of only SOM ($\delta^{13}C_{SOM}$) (in controls) and assuming that CO_2 produced by oil mineralization inherits its isotope composition ($\delta^{13}C_{oil}$), respectively, the share of CO_2 formed by mineralization of SOM (F_{SOM}) in experiments was calculated by expression [8].

$$F_{SOM} = (\delta^{13}C_{tot} - \delta^{13}C_{oil}) / (\delta^{13}C_{SOM} - \delta^{13}C_{oil}) \quad (8)$$

2.5 Cumulative CO_2 resulted from hydrocarbon mineralization

Cumulative CO_2 produced during the microbial substrate oxidation was calculated as follows. The ΔQ_i quantity of CO_2 evolved during the Δt_i-time interval (i = 1,2, …,n) was estimated as $\Delta Q_i = \Delta t_i \cdot v_i$, where the v_i-value is the rate of CO_2 evolved during the time interval Δt_i. Using $\delta^{13}C_{soil}$, $\delta^{13}C_{Subst}$ and $\delta^{13}C_{CO2(mix)(i)}$, the fraction of CO_2 resulting from the exogenous substrate (crude oil hydrocarbons) oxidation during Δt_i can be calculated as [9]:

$$\Delta Q_{Subst(i)} = (1 - F_{SOM(i)}) \cdot \Delta Q_i \quad (9)$$

where $F_{SOM(i)}$ value can be estimated using equation [8]. The cumulative CO_2 quantity ($Q_{Subst(CO2)}$) resulting from microbial oxidation of the substrates in soils was presented by [10], where *i* varied from 1 to *n*:

$$Q_{Subst(CO2)} = \Sigma\, \Delta Q_{Subst(i)} \quad (10)$$

2.6 Calculation of priming effects

The addition of exogenous test substrate (oil hydrocarbons) to soil was accompanied by the change in soil microbiota activity: the rate of CO_2 production initially increased as a result of substrate and probably SOM mineralization and then, on depletion of the substrate, gradually decreased. The amount of CO_2 evolved was divided by means of mass isotope balance into two fractions: from the substrates (oil hydrocarbons) and from SOM mineralization. Thus, the difference between CO_2 evolved from SOM mineralization in oil hydrocarbons amended soil (C_{*SOM}) and in the control soil (C_{SOM}) relative to the control (in percentage) was used to estimate the magnitude of the priming effect (PE) induced by oil hydrocarbons (denoted as SUB). The PE value was determined in two notations as *kinetic* PE(Δt_i) calculated as a value for Δt_i–time intervals using equation [11] and the PE(*total*) calculated as a weighted average value for the whole period of observation using equation [12].

$$PE(\Delta t_i)\ [\%] = 100\times(C_{*SOM(i)} - C_{SOM(i)})/C_{SOM(i)} \qquad (11)$$

where $C_{*SOM(i)} = F_i\times C_{(SUB+SOM)I}$; $C_{(SUB+SOM)i}$ is the total C evolved as CO_2 in the amended soil during Δt_i-time; and F_i is the share of CO_2-C resulting from the SOM in crude oil amended soil in Δt_i-time, which was calculated by equation [8].

$$PE(total)\ [\%]=\Sigma(PE(\Delta t_i)\cdot\Delta t_i)/\Sigma(\Delta t_i) \qquad (12)$$

where PE(Δt_i) was calculated according to Eq. [11].

3. Degradation of oil hydrocarbons by soil microbiota and laboratory bacteria introduced into soil

3.1 Soil samples

Arable soil samples from the Krasnodar region of Russia were used in the experiment after they had been cultivated with corn (C_4-plant). Soil samples were sieved through a 2 mm sieve and then moistened to 60 % of field capacity. The initial organic matter content was about 4.9 % of dry soil (DS) weight or 19.6 mg C g^{-1} DS. The carbon isotope composition in the initial SOM was characterized by a $\delta^{13}C$ value of -23.01 ± 0.2 ‰, typical of soils vegetated by C_4-plants.

3.2 Crude oil test-substrate

The crude oil as hydrophobic compound was applied as follows: crude oil (4 ml of oil corresponding to 3200 mg) was added to 10 g of dried and dispersed soil and then 10 g of the soil was mixed with fresh moist soil equivalent to 100 g of dry material. The final substrate concentration was 27.43 mg C g^{-1} soil. Since the content of SOM in the initial dry soil sample was about 19.6 mg C/g DS, the share of oil hydrocarbons introduced into the soil exceeded 1.4-fold the quantity of SOM. Assuming that the major part of crude oil spilled over the soil is contained in the upper 10-cm layer, we find that the supposed degree of soil pollution will be about 32 tons per 1 ha.

The carbon isotope composition of the oil hydrocarbons used in these experiments was characterized by a $\delta^{13}C$ value of -28.4 ± 0.1 ‰, the light and heavy oil hydrocarbon fractions having values -28.9 %o and -27.2 %o, respectively. The isotopic characteristics ($\delta^{13}C$) of the oil used in the experiments were found to be close to the samples of crude oil from oilfields of the Arabian region, where the $\delta^{13}C$ value was -27.5 ± 0.5 ‰ for oil, -28 ± 0.5 ‰ for alkane fraction, and -26.5 ± 1.5 ‰ for the fraction containing mainly aromatic hydrocarbons, respectively (Belhaj et al. 2002).

3.3 Microorganisms

To estimate the potential of microbial mineralization of oil hydrocarbons polluted soils, the CO_2 production was determined in 12 glass vials with tested soils (three replicates of each experiment and control) (Table 1). In Experiment 1, crude oil was introduced into vials with native soil containing only native soil microorganisms; in Experiment 2, the laboratory strain *Pseudomonas aureofaciens* BS1393(pBS216) (Kochetkov et al. 1997) was additionally introduced into the same soil with oil. Native soil without oil and the same soil with the strain BS1393(pBS216) were used as controls 1 and 2, respectively (Table 1).

The strain *Pseudomonas aureofaciens* BS1393(pBS216) bears the plasmid pBS216 that controls naphthalene and salicylate biodegradation, is able to utilize aromatic oil hydrocarbons, and has an antagonistic effect on a wide range of phytopathogenic fungi (Kochetkov et al. 1997). The ability of the strain to synthesize phenazine antibiotics and thus staining its colonies bright-orange on LB agar medium allowed its use as a marker of quantitative presence of the above microorganisms in soil in the presence of aboriginal microflora (Sambrook, et al. 1989].

Control 1: Native soil with soil microbiota (three of glass vials)	*Control* 2: Native soil with soil microbiota + *Pseudomonas aureofaciens* BS1393(pBS216) (three of glass vials)
Experiment 1: Native soil with soil microbiota + crude oil (three of glass vials)	*Experiment* 2: Native soil with soil microbiota + crude oil+ *Pseudomonas aureofaciens* BS1393(pBS216) (three of glass vials)

Table 1. Scheme of experiments and controls

The introduced strain was previously grown in liquid LB medium till stationary phase (28°C, 18 h) and then uniformly introduced into soil to a concentration of 10^6 cells g^{-1} soil. The control of the bacteria strain growth was accomplished weekly during 67 days. A composite soil sample was collected from three separate sub-samples from the vial and analyzed for bacterial quantities. Approximately one g of the composite soil sample was suspended in 10 ml of 0.85% NaCl on "Vortex", soil particles were precipitated, and 1 ml of supernatant was used for making dilutions (10×-10000×). Volume of 0.1 ml of the corresponding dilutions was inoculated onto Petri dishes with LB medium. The colony-forming units (CFU) on the plates were counted and their mean values in the control and experiments were calculated.

As seen from Table 2, in one day after introduction of the strain *P. aureofaciens* BS1393(pBS216) experiments (soil with oil) and controls (soil without oil) showed a decrease of the quantity of cells of this strain from 10^6 cells g^{-1} soil to 10^4 cells g^{-1} soil measured as colony-forming units (CFU). However, in 7 days after the beginning of the experiment, the CFU number of the bacteria introduced in the experiment with oil was about 2.7 $\times 10^6$ cells g^{-1} DS, i.e. more than 17-fold higher than the CFU of the same bacterium in the control soil without oil (Table 2). These results indicate the ability of the strain introduced for biodegradation of oil hydrocarbons to utilize them as a growth substrate. In 14-21 days, the CFU of the introduced strain noticeably decreased again and by day 28 reached the initial level of 10^4 cells g^{-1} DS.

Variants	Colony-forming units ($x10^4$)/g of soil					
	*1 d	7 d	14 d	21 d	28 d	35 d
Control	8.0 (1.7)	15.7 (5.8)	4.0 (0.9)	1.6 (0.6)	3.1 (3.5)	2.2 (0.9)
Experiment	4.7 (3.4)	268.0 (149)	31.7 (18.7)	21.1 (14.4)	2.4 (1.4)	1.5 (0.5)

*Times after bacteria culture was introduced into soil. (Standard deviations from 3 parallels are given in parenthesis)

Table 2. Growth of *Pseudomonas aureofaciens* BS1393(pBS216) without (control) and with crude oil hydrocarbons (experiment) to a concentration of 10^6 colony-forming units g^{-1} of soil introduced into arable soil.

3.4 Microbial CO_2 production in soil

Total rates of microbial mineralization of SOM and oil hydrocarbons in soil were determined by the rate of CO_2 production (µg C-CO_2 g^{-1} DS h^{-1}).

In controls 1 and 2, the rates of SOM mineralization both by aboriginal soil microorganisms and the mixture of these microorganisms plus introduced strain *P. aureofaciens* BS1393(pBS216) were within the range of 0.2 ± 0.02 µg C-CO_2 g^{-1} DS h^{-1} and practically did not change during the 67-day observation (Fig. 1, control 1 and control 2). In soil with added oil hydrocarbons (experiments 1 and 2), the rate of mineralization of total organic carbon significantly increased and reached the maximum value of about 3.2 µg C-CO_2 g^{-1} DS h^{-1} on days 7-8 after the beginning of the exposure (Fig. 1, Exp. 1 and Exp. 2). In experiment 2, with the bacterium P. *aureofaciens* BS1393(pBS216) added to the indigenous microbiota, there are two maximums of CO_2 production rate: the first in 3 days and the second one in 8 days after the beginning of exposure (Fig. 1, Exp. 2). In the experiment 1 with aboriginal microbiota (Fig. 1, Exp. 1) only one maximum of CO_2 production rate was observed in 7 days after the beginning of exposure. It is supposed that this special feature was responsible for the availability of the introduced bacteria P. *aureofaciens* BS1393(pBS216) to consume the oil hydrocarbons.

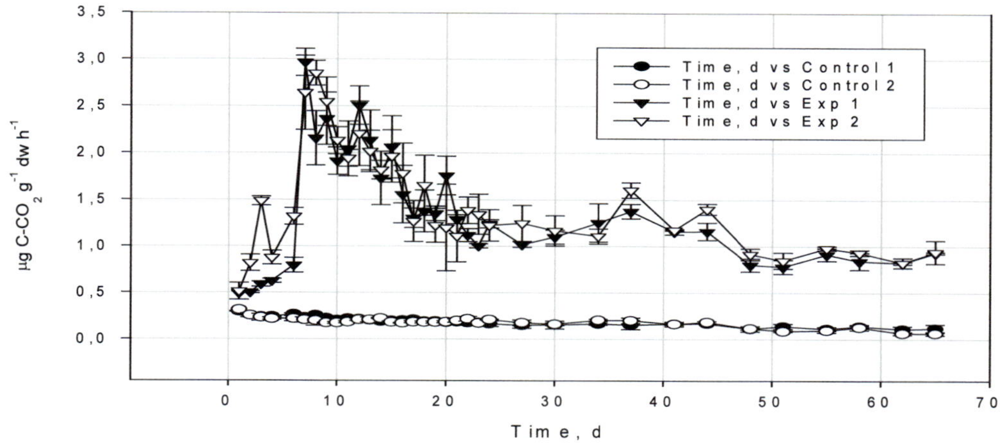

Fig. 1. Rates of CO_2 production by microbial mineralization of substrates in experiments simulating microbial utilization of oil hydrocarbons. Control 1 (aboriginal microflora); Control 2 (aboriginal microflora + introduced bacteria); Experiment 1 (aboriginal microflora + oil); Experiment 2 (aboriginal microflora + introduced bacteria + oil)

Two to 3 days (Exp. 2) and 5 to 6 days (Exp. 1) days after the start of exposure, the crude oil introduced into agricultural soil caused an exponential increase in the CO_2 emission rate indicating microbial growth after lag-phase (Fig. 2).

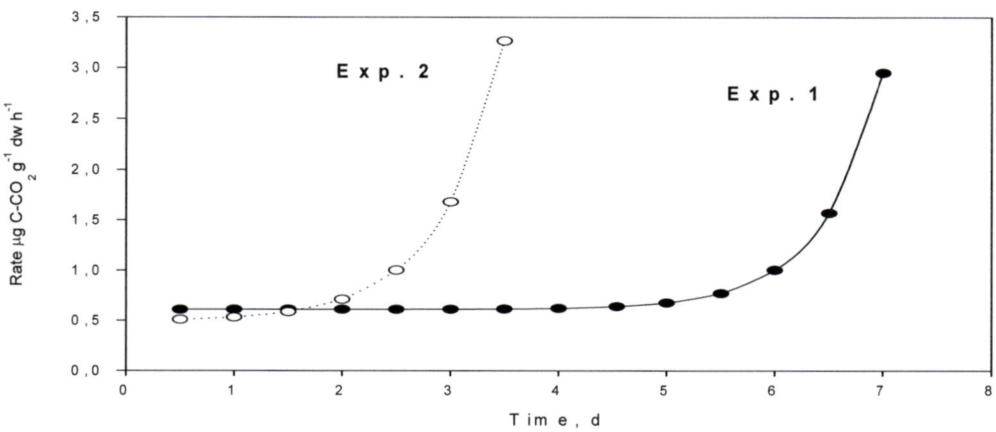

Fig. 2 Substrate-induced respiratory response of the microbial community during incubation of soil treated with crude oil hydrocarbons: 1 - the initial CO_2 emission by growth of native soil microbiota and 2- the initial CO_2 emission by growth of mixture of native soil microbiota with strain *P. aureofaciens* BS1393(pBS216)

At the initial stages of microbial oil mineralization in experiments 1 and 2, the specific rates of metabolic CO_2 emission (μ) were determined using the approximating equation [1] and lag periods (t_{lag}) were calculated by the equation [2] (Table 3). The values of parameters K as an index of catabolism of microbial cells in soil were calculated from the analysis of CO_2 production at the initial stages of microbial oil mineralization. The values of parameters K (Table 3) show the close rates of initial production of metabolic CO_2 in these experiments. At the same time, parameter *r* indicating the presence of growing microorganisms in soil is higher by three orders of magnitude in experiment 2 with introduced bacteria compared to experiment 1 with native microbiota in soil. Parameter μ showing specific rates of CO_2 production in experiments 1 and 2 has close values within the measurement error. As one would expect, the lag period of test substrate consumption and CO_2 production in experiment 2 with the introduced bacterium P. *aureofaciens* BS1393(pBS216) was about 2,5±03 days, i.e., significantly less than in experiment 1 with native microbiota only (the lag period of 6,2±0,5 days).

Type of soil	μg CO_2-C g^{-1} soil h^{-1}			t_{lag}, d
	K	r	μ	
Native soil microbiota (Experiment. 1)				
Agricultural soil	0.6085	$8.991 \cdot 10^{-6}$	1.7814	6.2 (0.5)
Native soil microbiota + *P. aureofaciens* BS1393(pBS216) (Experement. 2)				
Agricultural soil	0.4906	$7.445 \cdot 10^{-3}$	1. 6913	2.5 (0.3)

Table 3. Parameters of the equations [1] and [2] characterized the respiration rates of native soil microbiota (Experiment 1) and mixture microbiota after bioagmentation with strain *P. aureofaciens* BS1393(pBS216) (Experiment 2) after crude oil addition to the agricultural soil. Standard deviation intervals are in brackets

Beginning from day 25 to day 67 from exposure, the rate of CO_2 production in experiments 1 and 2 decreased slightly and stabilized at a level of 1.25 ± 0.25 μg C-CO_2 g^{-1} DS h^{-1} (Fig. 1). Total CO_2 production in controls (control 1 and 2) for the 47-day and for 67-day periods of observation was 24.8 ±1.2 mg C-CO_2 and 35.5 ± 1.2 mg C-CO_2 (Table 4).

Experiment	Mean Production rate, μg C-CO2 g-1 DS h-1	*Total production, mg C-CO2	**Time, days
Control 1	0.228(0.013)	25.7 (0.6)	47
Control 1	0.228(0.013)	36.7 (0.6)	67
Control 2	0.213(0.013)	24.03 (0.6)	47
Control 2	0.213(0.013)	34.25 (0.6)	67
Experiment 1	1.480(0.122)	167 (6)	47
Experiment 1	1.480(0.122)	238 (6)	67
Experiment 2	1.546(0.100)	174 (5)	47
Experiment 2	1.546(0.100)	251 (5)	67

*Total production Q_{total}=(24 $\cdot v_{average}$ (μg C-CO_2 g^{-1} DS h^{-1}) · t (days))x100 g DS

**Time after the crude oil addition to soil. Standard deviations of three parallel determinations are given in brackets.

Table 4. Mean rates of CO_2 emission (μg C-CO_2 g^{-1} DS per h) and total production of C-CO_2 during the time experiment (mg C-CO_2 per 100 g DS)

The absence of any significant differences in CO_2 production in controls 1 and 2 was considered as an evidence of insignificant additional mineralization of SOM attributable to the introduced strain of *P. aureofaciens* BS1393(pBS216). In the case of oil-containing soils, the amounts of metabolic CO_2 in experiments 1 and 2 exceeded 6.8-fold that of controls 1 and 2, being 167.0 and 238 mg C-CO_2 (Exp. 1) and 174.0 and 251 mg C-CO_2 (Exp. 2) during 47- and 67-day exposure, respectively (Table 4). The data also showed that the additional introduction of the hydrocarbon-oxidizing strain *P. aureofaciens* BS1393(pBS216) into oil-containing soil (Exp. 2) promoted the increase of metabolic CO_2 amount (4 - 13 %) compared to the aboriginal soil microorganisms.

Total CO_2 production in experiments 1 and 2 included microbial mineralization of SOM and oil hydrocarbons, therefore the share of CO_2 formed by mineralization of each of the above substrates was determined by measuring values $\delta^{13}C$, both in the carbon isotope characteristics of SOM and oil products and in the metabolic carbon dioxide formed during this process.

3.5 Analysis of the origin of soil CO_2 using $^{13}C/^{12}C$ ratios

In experiments 1 and 2, the $\delta^{13}C$ values of the metabolic CO_2 released from soil in the 3 days before oil hydrocarbons introduction into soil were -23.53 ± 0.21 %o and -23.56 ± 0.25 %o, respectively, and actually identical to the isotopic characteristics of CO_2 in the controls (Fig. 3).

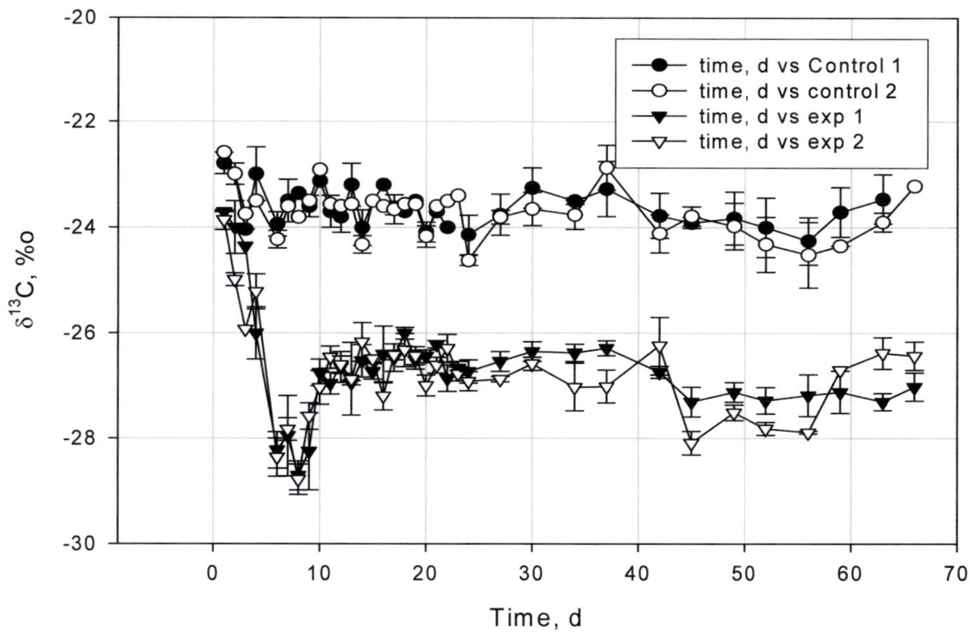

Fig. 3. Carbon isotope characteristics ($\delta^{13}C$, ‰) of CO_2 produced in experiments of microbial mineralization of SOM and oil hydrocarbons introduced into soil: Control 1 (aboriginal microflora); Control 2 (aboriginal microflora + introduced bacteria); Experiment 1 (aboriginal microflora + oil); Experiment 2 (aboriginal microflora + introduced bacteria + oil)

After oil hydrocarbons addition to soil, the share of ^{13}C isotope in metabolic carbon dioxide abruptly dropped, which was an evidence of CO_2 production partly from oil hydrocarbons containing less ^{13}C isotope compared to SOM. The maximum depletion of ^{13}C isotope in metabolic CO_2 was registered during the days 11-15 from the beginning of exposure in experiments 1 and 2. This was considered a result of the mineralization of mainly alkane oil fractions. Our assumption that the major part of aliphatic hydrocarbons from the introduced crude oil had already been utilized by that period is evidenced by the carbon isotope characteristics of the metabolic carbon dioxide with the value of $\delta^{13}C = -28.5 \pm 0.5$ %o (Fig. 3). After 15 days and until the end of the experiment (67 days), the isotopic characteristic of CO_2 was at around the value of $\delta^{13}C = -26.8 \pm 0.5$ %o. Using equation [4], the average weighted isotope composition of CO_2 produced by microbial mineralization of total organic products (oil hydrocarbons and SOM) in experiments 1 and 2 during 67-days was characterized by $\delta^{13}C$ values about of -26.6 ± 0.1 %o, which significantly differed from the carbon isotope characteristics of oil ($\delta^{13}C = -28.4 \pm 0.2$%o) and SOM ($\delta^{13}C = -23.01 \pm 0.2$ ‰,). It can be said with confidence that metabolic CO_2 was produced during microbial mineralization of a part of SOM and a part of oil hydrocarbons.

3.6 Priming effect of oil hydrocarbons

The *kinetic* PE was calculated by comparing CO_2 amounts generated by microbial mineralization of SOM and oil products (Exp. 1 and 2) to CO_2 amounts generated in the controls in the corresponding periods of observation [Eq. 11].

In order to quantify both the extent and direction of PE of oil hydrocarbons, we have compared the rates of CO_2 production by microbial mineralization of SOM before and after introduction of oil hydrocarbons into soil at the initial period of maximum microbial activity, i.e., during 15 days after addition of crude oil to soil (Fig. 4). As shown in Figure 4 (A), the activation of the metabolism of aboriginal hydrocarbon-oxidizing soil microorganisms in experiments 1 took about 6 days from the introduction of the hydrocarbon substrate, when microbial rate of CO_2 production increased to a rate closer to that of experiment 2 with the *P. aureofaciens* BS1393(pBS216) addition. The mass isotope balance data showed that during these days in experiment 1 the mineralization of oil hydrocarbons was insignificant and the rate of SOM mineralization was less the rate in control (negative PE) (Fig. 4, C PE_1). Experiment 2, in contrast to experiment 1, showed the utilization of oil hydrocarbons in the initial period of exposure was accompanied by a noticeable increase of SOM mineralization rate compared to the initial value (positive PE) (Fig. 4, C PE_2). However, PE became negligible in both experiments during 6-8 day exposure; it is possibly the mineralization time of aliphatic hydrocarbons or their partially oxidized products. The negative PE has been demonstrated previously in the processes of the microbial mineralization of n-hexadecanoic acid introduced into soil (Zyakun et al. 2011). At the next period of the exposure, the PE values demonstrate the positive values of 300 % in experiment 1 and about 400 % in experiment 2. On completion experiments, the total PE has been calculated using Eq. {12}. Taking into account the CO_2 quantity registered in experiments 1 and 2 during the whole period of exposure (Table 4, Q_{total}) and the share of CO_2 under microbial utilization of SOM (Table 5, F_{SOM}), we find the quantity of CO_2 form as a result of SOM mineralization in the experiments (Table 5, $[CO_2]_{SOM}$)

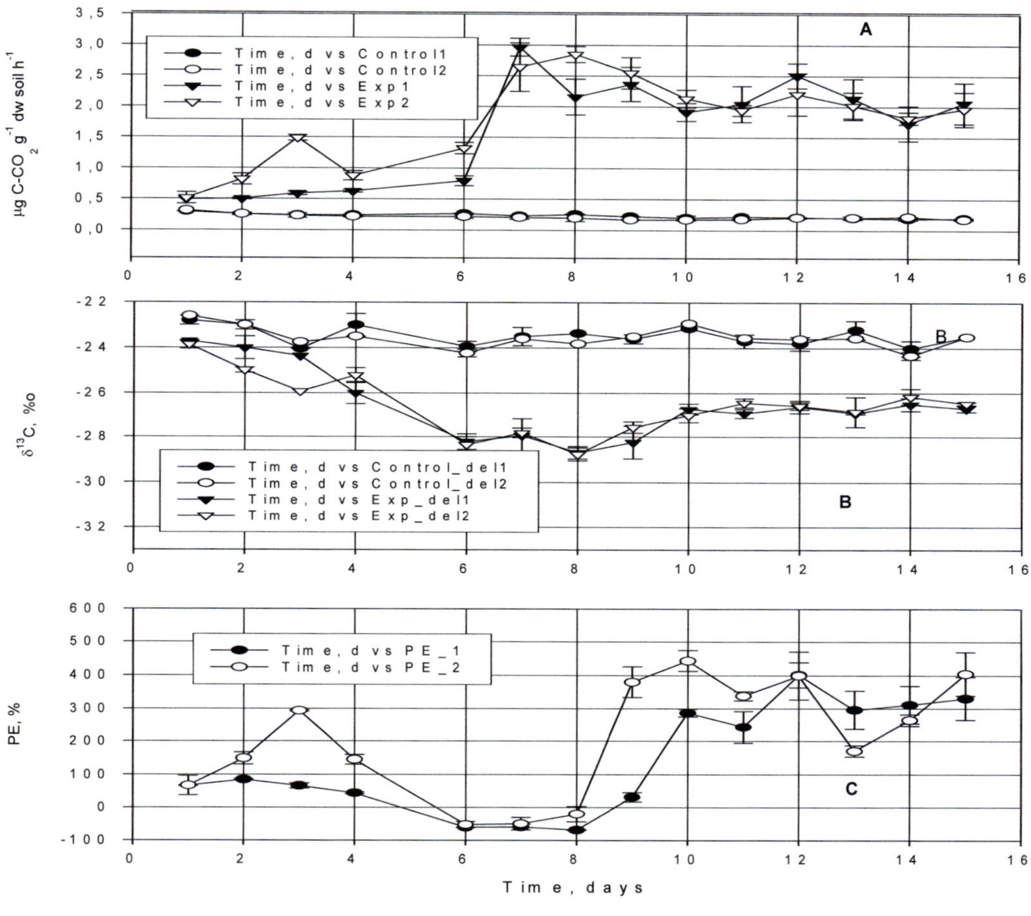

Fig. 4. Rates (A) and isotopic characteristics of CO_2 resulting from SOM and oil hydrocarbons mineralization (B), and priming effect (C) for 15 days after oil introduction into soil: control 1 (aboriginal microflora); control 2 (aboriginal microflora + introduced bacteria); experiment 1 (aboriginal microflora + oil); experiment 2 (aboriginal microflora + introduced bacteria + oil).

Experiment	*$\delta^{13}C_{ave}$, %o	**F_{SOM}, %	$[CO_2]$(SOM) mg C-CO_2	#PE, %	##Time, days
Control 1	-23.70 (0.1)	100	25.7 (0.6)	0	47
Control 1	-23.70 (0.1)	100	36.7 (0.6)	0	67
Control 2	-23.77 (0.1)	100	24.03 (0.6)	0	47
Control 2	-23.77 (0.1)	100	34.25 (0.6)	0	67
Experiment 1	-26.59 (0.2)	38.5 (1.7)	64 (3)	150 (13)	47
Experiment 1	-26.59 (0.2)	38.5 (1.7)	92 (3)	151 (13)	67
Experiment 2	-26.63 (0.2)	38.2 (1.6)	67 (3)	177(15)	47
Experiment 2	-26.63 (0.2)	38.2 (1.6)	96 (3)	180 (15)	67

*$\delta^{13}C_{ave}$ is an average weighted of isotope characteristic of CO_2 was calculated [Eq. 4]
**F_{SOM} is a share of metabolic CO_2 formed by microbial mineralization of SOM; # PE is a priming effect was calculated according to [Eq. 11]; ##Time after the crude oil addition to soil. Standard errors of three parallel calculations are given in brackets.

Table 5. Average weighted characteristics ($\delta^{13}C_{ave}$) of carbon isotope composition and fraction of CO_2 formed by SOM mineralization and priming effect (PE) in experiments 1 and 2 relative to controls

Using the equation [12], we calculate the value of PE(total) by comparing CO_2 production during microbial SOM utilization in the experiments and controls. As follows from Table 5, during 67-day exposure of oil hydrocarbons in soil the PE value reached 150 % in experiment 1 with native soil microbiota and 180 % in experiment 2 with mixed microbiota (soil microorganisms and the bacterium strain *P. aureofaciens* BS1393(pBS216)). Thus, addition of crude oil to the soil activates to a large extent the microbial mineralization of native soil organic matter.

3.7 Microbial utilization of oil hydrocarbons and SOM transformation

As follows from Table 6, the oil hydrocarbons introduced into soil were mineralized to CO_2 to the extent of about 4.59 (0.2) and 4.81 (0.15) mg C-CO_2 g-1 DS or 16.7 and 17.5 percents of the initial crude oil quantities in the soil over the course of 67-day exposure in experiments 1 and 2, respectively.

Variants of analysis	Initial C_{org}, (SOM + Oil) mg C g^{-1} DS	aC-SOM mineralized, mg C-CO_2 g^{-1} DS	Crude oil metabolized C_{oil}, mg C g^{-1} DS			bR
			CO_2	Biomass	cTotal	
Experiment 1	19.6+ 27.43	2.87(0.2)	4.59 (0.2)	4.59 (0.2)	9.18 (0.2)	1.60
		c14.6 %	d16.7 %	d16.7 %	33.4 %	
Experiment 2	19.6+ 27.43	2.98(0.15)	4.81 (0.15)	4.81 (0.15)	9.62 (0.15)	1.61
		15.2 %	17.5 %	17.5 %	35.0 %	

*The CO_2 evaluation from SOM calculated as $Q_{CO2(SOM)} = v_{CO2(SOM+SUB)} \cdot \Delta t \cdot F_{SOM}$
bR= (Q biomass + exometabolites from oil carbon) / (Q_{SOM} mineralized of SOM);
cParts (%) of the initial amount of SOM and crude oil mineralized to CO_2 in soil dParts of the initial amount of crude oil (in percents) consumed by microorganisms producing CO_2 and organic substances (biomass and exometabolites). Standard deviations are given in brackets.

Table 6. The quantities of SOM mineralization and crude oil consumption by microbiota during the 67-day exposure in soil.

Previously (Zyakun et al. 2003), it was shown that during the growth of microbial cells on hydrocarbons the ratio of biomass and CO_2 carbon quantities was corresponding 1:1. In view of the above, we believe that the quantity of oil hydrocarbons taken up for the biosynthesis of cell biomass and organic exometabolites in soil during the 67-day exposure will be close to the carbon quantity of CO_2 production and make no less then 16.7 and 17.5 percents of the oil introduced in experiments 1 and 2, respectively. By this is meant that the oil hydrocarbon consumption by microbial pool in soil amounts up 33.4 and 35 percent of total oil, respectively.

Extrapolation of the obtained data (Table 6) to a 6-month season, when the temperature conditions in the Krasnodar region provide for the metabolic activity of soil microbiota, shows that the uptake of crude oil hydrocarbons by native soil microbiota may reach no more than 92±2 % of the total oil hydrocarbon quantity in the oil. At a positive PE of oil hydrocarbons in soil, there is more intensive microbial degradation of SOM compared to the processes in native soil. On the other hand, oil hydrocarbons consumed by microorganisms are spent both for CO_2 production and for the biosynthesis of biomass and organic exometabolites, which then are included in SOM and transform the structure of soil. The newly synthesized metabolites and microbial biomass components can be used by other biological systems (plants, macro- and microorganisms) that are incapable of direct utilization of oil hydrocarbons. The quantitative and isotopic data obtained in the experiments were used as a basis for estimation of the degree of replacement of part of SOM mineralized to CO_2 by the newly synthesized products under microbial utilization of oil hydrocarbons. Table 6 shows the rates of microbial degradation and production of cell biomass and organic exometabolites in model experiments with microbial utilization of crude oil as a substrate. As a result of oil consumption both by native soil microbiota (Exp. 1) and introduced the bacterium strain *P. aureofaciens* BS1393(pBS216) (Exp. 2), the quantity of the newly synthesized organic products (carbon of cell biomass and exometabolites) nearly 1.6-fold exceeds the carbon quantity of SOM taken up for the CO_2 mineralization (Table 6, R). It means that microbial transformation of oil hydrocarbons into products available as substrates for other living systems may be a peculiar source of organic fertilizers. In addition, there is more and more evidence that the bioremediation of oil-polluted soils is companied by plant growth stimulation.

4. Conclusion

With the proviso that crude oil carbon content no more than 1.4-fold higher than the SOM carbon amount, the soil microbiota is able to mineralize up to 17 % of crude oil hydrocarbons and 15 % of SOM during the 67-day experiments. Using mass isotope balance and differences between the $\delta^{13}C$ values of SOM and oil hydrocarbons, the quantities of CO_2 produced during microbial mineralization of SOM and oil hydrocarbons have been determined. According to the highest depletion of ^{13}C in CO_2 evolved from soil during the initial time of the exposure with crude oil, it is suggested that at this time the aliphatic oil fraction predominantly participates in mineralization. Microbial consumption of oil hydrocarbons activates the process of SOM mineralization and demonstrates the presence of PE of oil hydrocarbons. During a 67-day period of the crude oil exposure in soil, the average values of PE reached over 150 % in soil with native soil microbiota and over 180 % in soil with the mixture of native microbiota and introduced bacteria *P. aureofaciens*

BS1393(pBS216) containing the plasmid pBS216 which controls naphthalene and salicylate biodegradation and able to utilize aromatic oil hydrocarbons. It has been found experimentally that in the total emission of carbon dioxide from soil to atmosphere, about 38 % CO_2 was produced as a result of SOM mineralization and about 62 % was formed from oil hydrocarbons as anthropogenic pollutant. The soils polluted with oil hydrocarbons undergo the change of SOM by replacement of part native organic substances on the newly synthesized products in the course of oil biodegradation and the increase of the residual oil share in the total pool of organic matter in soil. Within 6-month time, the quantity of the microbial newly synthesized organic products (carbon of cell biomass and exometabolites) nearly 1.6-fold exceeds the carbon quantity of SOM taken up for the CO_2 microbial mineralization. After partially microbial consumption of oil hydrocarbons, the substrate characteristics of residual oil are rather different from crude oil and can be considered as *waste oil* in the soil.

5. References

Abbassi B.E., Shquirat, W.D. (2008). Kinetics of indigenous isolated bacteria used for ex-situ bioremediation of petroleum contaminated soil. *Water Air Soil Pollution,* Vol. 192, pp. 221–226

Adam G., Duncan H. (2002). Influence of diesel fuel on seed germination. *Environmental Pollution,* Vol. 120, pp. 363-370.

Adam G., Duncan H. (2003). The effect of diesel fuel on common vetch (*Vicia sativa* L.) plants. *Environmental Geochemistry and Health,* Vol. 25, pp. 123-130.

Anderson J.P.E., and Domsch K.H. (1978). A physiological method for the quantitative measurement of microbial biomass in soils. *Soil Biol. Biochem..* Vol. 10, pp. 215-221

Belhaj A., Desnoues N. and Elmerich C. (2002). Alkane biodegradation in *Pseudomonas aeruginosa* strains isolated from pullulated zone: identification of *alk*B and *alk*B-relative genes. //*Res. Microbiol.* Vol. 153. P. 339–344.

Bingeman C.W., Varner J.E., and Martin W.P. (1953). The effect of the addition of organic materials on the decomposition of an organic soil. *Soil Sci. Soc.* Vol. 22, pp. 707-713.

Blagodatskaya, E.V., Blagodatsky, S.A., Anderson, T.-H., Kuzyakov Y. (2007). Priming effects in Chernozem induced by glucose and N in relation to microbial growth strategies. *Appl. Soil Ecology,* Vol. 37, pp. 95-105.

Blagodatskaya, E.V., Kuzyakov Y. (2008). Mechanism of real and apparent priming effects and theiir dependence onsoil microbial biomass and community structure: critical review. *Biol. Fertil. Soils,* Vol. 45, pp. 115-131.

Blagodatsky S.A., Heinemeiyer O. Richter J. (2000). Estimating the active and total soil microbial biomass by kinetic respiration analysis. *Biol. Fert. Soils.* Vol. 32, pp. 73-81.

Blagodatskaya, E.V., Blagodatsky, S.A., Anderson, T.-H., Kuzyakov Y. (2009). Contrasting effects of glucose, living roots and maize straw on microbial growth kinetics and substrate availability in soil. *European J. Soil Sci.* Vol. 60, pp. 186-197

Craig H. (1957). Isotopic standards for carbon and oxygen and correction factors for mass-spectrometric analysis of carbon dioxide. *Geochim. Cosmochim. Acta,* Vol. 12, pp. 133-140.

Hamamura N., Olson S.H., Ward D. M., and Inskeep W.P. (2006). Microbial population dynamics associated with crude-oil biodegradation in diverse soils. *Appl. Environ. Microbiol.* Vol. 72, No. 9, pp. 6316-6324

Hamamura N., Fukui M., Ward D.M., Inskeep W.P. (2008). Assessing soil microbial populations responding to crude-oil amendment at different temperatures using phylogenetic, functional gene (alkB) and physiological analyses. *Environ. Sci. Technol.* Vol. 42, No. 20, pp. 7580-7586;

Harabi N.E-D. and Bartha R. (1993).Testing of some assumptions about biodegradability in soil as measured by carbon dioxide evolution. *Appl. & Environ. Microbiol.* V. 59. No. 4, pp. 1201-1205.

Jobson A., McLaughlin M., Cook F. D., Westlake D.W. S. (1974). Effect of amendments on the microbial utilization of oil applied to soil. *Appl. Microbiol.* Vol. 27, No. 1, pp. 166-171.

Juck D, Charles T., White L., Greer C. (2000). Polyphasic microbial community analysis of petroleum hydrocarbon-contaminated soils from two northern Canadian communities. *FEMS Microbiol. Ecol.* Vol. 33, pp. 241-249.

Kaplan C.W., Kitts C.L. (2004). Bacterial succession in petroleum land treatment unit. *Appl. Environ. Microbiol.* Vol. 70, pp. 1777-1786.

Khan R.A., and Ryan P. (1991). Long Term Effects of Crude Oil on Common Murres (*Uria Eagle*) Following Rehabilitation .*Bulletin of Environmental Contamination and Toxicology,* Vol. 46, No 2, pp. 216-222.

Kochetkov V.V., Balakshina V.V., Mordukhova E.A., Boronin A.M. (1997). Plasmids of naphthalene biodegradation in rhizosphere *Pseudomonas* bacteria. *Microbiology* (in Russian). V. 66, No.2, pp. 211 - 216.

Margesin R., Schinner F. (2001). Bioremediation (natural attenuation and biostimulation) of diesel-oil-contaminated soil in an alpine glacier skiing area. *Appl. Environ. Microbiol.* Vol. 67, No. 7, pp. 3127-3133.

Margesin R., Hämmerle M., Tscherko D. (2007). Microbial activity and community composition during bioremediation of diesel-oil-contaminated soil: effects of hydrocarbon concentration, fertilizers, and incubation time. *Microbial Ecology,* Vol. 53, pp. 259-269.

Margesin, R., and Schinner, F. (eds), (2005). Mannual for soli analysis-monitoring and assessing soil bioremediation. Soil Biology, Vol. 5, Springer Verlag, Berlin, 359 pp.

Mishra S., Jyot J., Kuhad R.C., Lal B. (2001). Evaluation of inoculum addition to stimulate in situ bioremediation of oily-sluge-contaminated soil. *Appl. Environ. Microbiol.* Vol. 67, pp. 1675-1681.

Mundi I., (2010). "World Crude Oil Consumption by Year",. http://www.indexmundi.com/energy.aspx].

Nikitina E.V., Yakusheva O.I., Zaripov S.A., Galiev R.A., Garusov A.V., Naumova R.P. (2003). Distribution and physiological state of microorganisms in petrochemical oily sludge. *Microbiology.* Vol. 72, pp. 621–627.

Palmroth M.R. Munster U., Pichtel J., Puhakka J.A. (2005). Metabolic response of microbiota to diesel fuel addition in vegetated soil. *Biodegradation.* Vol. 16, pp. 91-101.

Panikov N.S. (1995). *Microbial Growth Kinetics.* Chapman and Hall, London, Glasgow, 378 pp.

Panikov N.S., Sizova M.V. (1996). A kinetic method for estimating the biomass of microbial functional groups in soil. *Journal of Microbiological Methods.* Vol. 24, pp. 219-230.

Pleshakova E.V., Dubrovskaya E.V., Turkovskaya O.V. (2008). Efficiencies of introduction of an oil-oxidizing *Dietzia maris* strain and stimulation of natural microbial communities in remediation of polluted soil. *Applied Biochemistry and Microbiology.* Vol. 44, No. 4, pp. 389–395.

Sambrook, J., Maniatis, T. and Fritsch, T.F. (1989). Molecular cloning: A Laboratort Manual. Cold Spring Harbor Laboratory Press, Cold Spring Harbor, NY.

Shen J. and Bartha R. (1996). Priming effect of substrate addition in soil-based biodegradation tests. *Appl. & Environ. Microbiol.* V. 62, No. 4, pp. 1428-1430.

Stenström J, Stedberg B., Johanson M. (1998). Kinetic of substrate-induced respiration (SIR): Theory. *Ambio.* Vol. 27, No. 1, pp. 35-39;

Sikkema J., deBont A.M., and Poolman B. (1995). Mechanisms of Membrane Toxicity of Hydrocarbons. *Microbial Rev.* Vol. 59, pp. 201-222

Speight J.G. (1990). The Chemistry band technology of Petroleum, Marcel Dekker, New York,

Tevvors J.T., and Sair, M.H. Jr. (2010). The Legacy of Oil Spills. *Water, Air, and Soil Pollution.* Vol. 211, No 1, pp. 1-3. doi:10..1007/s11270-010-0527-5.

Tzing S.H., Chang J.Y., Ghule A., Chang J.J., Lo B., and Ling Y.C. (2003). A simple and rapid method for identifying the source of spilled oil using an electronic nose; conformation by gas chromatography with mass spectrometry. *Rapid Commun. Mass Spectrometry.* Vol. 17. No 16, pp. 1873-1880]

Van Hamme J.D., Singh A., Ward O.P. (2003). Recend advantages in petroleum microbiology. *Microbial Mol. Biol. Rev.* Vol. 67, No. 4, pp. 503-549.

Wang O., Zhang S., Li Y., Klassen W. 2011. Potential approaches to improving biodegradation of hydrocarbons for bioremediation of crude oil pollution. J. Environ. Protection, No 2, pp. 47-55. doi:10..4236/jep. 2011.21005,

Wongsa P., Tanaka M., Ueno A., Hasanuzzaman M., Yumoto I., and Okuyama H. (2004). Isolation and Characterization of Novel Strains of Pseudomonas aeruginosa and Serratia Marcescens Possessing High Efficiency to Degrade Gasoline, Kerosene, Diesel Oil and Lubricating Oil. *Current Microbiol.*. Vol. 49, pp. 415-422,

Zucchi M., Angiolini L., Borin S., Brusetti L., Dietrich N., Gigliotti C., Barbieri P., Sorlini C., Daffonchio D. (2003). Response of bacterial community during bioremediation of an oil-polluted soil. *J. Appl. Microbiol.* Vol. 94, pp. 248-257;

Zyakun A.M., Kosheleva, I.A., Zakharchenko, V.N., Kudryavtseva, A.I., Peshenko, V.P., Filonov, A.E., Boronin, A.M. (2003). The use of the [^{13}C]/[^{12}C] ratio for the assay of the microbial oxidation of hydrocarbons. *Microbiology.* Vol. 72, pp. 592-596

Zyakun A., Dilly O. (2005).Respiratory quotient and priming effect in an arable soil induced by glucose. *Appl. Biochem. and Microbiol.* Vol. 41. No 5. pp. 512-520

Zyakun A., Nii-Annang S., Franke G., Fischer T., Buegger F., and Dilly O. (2011). Microbial activity and ^{13}C/^{12}C ratio as evidance of n-hexadecane and n-hexadecanoic acid biodegradation in agricultural and forest soils. *Geomicrobiology J.* Vol. 28, pp. 632-647. doi: 10.1080/01490451.2010.489922

5

Earthworms and Vermiculture Biotechnology

A. A. Ansari[1,2] and S. A. Ismail[3]
[1]*Department of Biological Sciences, Faculty of Science*
[2]*Kebbi State University of Science and Technology*
[3]*Managing Director, Ecoscience Research Foundation*
[1,2]*Nigeria*
[3]*India*

1. Introduction

Earthworms are terrestrial invertebrates belonging to the Order Oligochaeta, Class Chaetopoda, Phylum Annelida, which have originated about 600 million years ago, during the pre-Cambrian era (Piearce *et al.*, 1990). Earthworms occur in diverse habitat, exhibiting effective activity, by bringing about physical and chemical changes in the soil leading to improvement in soil fertility. An approach towards good soil management, with an emphasis on the role of soil dwellers like earthworms, in soil fertility, is very important in maintaining balance in an ecosystem (Shuster *et al.*, 2000).

The role of earthworms in soil formation and soil fertility is well documented and recognised (Darwin, 1881; Edwards *et al.*, 1995; Kale, 1998; Lalitha *et al.*, 2000). The main activity of earthworms involves the ingestion of soil, mixing of different soil components and production of surface and sub surface castings thereby converting organic matter into soil humus (Jairajpuri, 1993). Earthworms play an important role in the decomposition of organic matter and soil metabolism through feeding, fragmentation, aeration, turnover and dispersion (Shuster *et al.*, 2000).

Earthworms were referred by Aristotle as "the intestines of earth and the restoring agents of soil fertility" (Shipley, 1970). They are biological indicators of soil quality (Ismail, 2005), as a good population of earthworms indicates the presence of a large population of bacteria, viruses, fungi, insects, spiders and other organisms and thus a healthy soil (Lachnicht and Hendrix, 2001).

The role of earthworms in the recycling of nutrients, soil structure, soil productivity and agriculture, and their application in environment and organic waste management is well understood (Edwards *et al.*, 1995; Tomlin *et al.*, 1995; Shuster *et al.*, 2000; Ansari and Ismail, 2001a, b; Ismail, 2005; Ansari and Ismail, 2008; Ansari and Sukhraj, 2010).

2. Ecological strategies of earthworms

Lee (1985), recognised three main ecological groups of earthworms, based on the soil horizons in which the earthworms were commonly found *i.e.*, litter, topsoil and sub soil.

Bouché (1971, 1977), also recognised three major groups based on ecological strategies: the epigeics (Épigés), anecics (Anéciques) and endogeics (Éndogés). Epigeic earthworms live on the soil surface and are litter feeders. Anecic earthworms are topsoil species, which predominantly form vertical burrows in the soil, feeding on the leaf litter mixed with the soil. Endogeic earthworms preferably make horizontal burrows and consume more soil than epigeic or anecic species, deriving their nourishment from humus.

2.1 Distribution of earthworms

Earthworms occur all over the world, but are rare in areas under constant snow and ice, mountain ranges and areas almost entirely lacking in soil and vegetation (Edwards and Bohlen, 1996). Some species are widely distributed, which are called peregrine, whereas others, that are not able to spread successfully to other areas, are termed as endemic (Edwards and Lofty, 1972).

2.2 Factors affecting earthworm distribution

The distribution of earthworms in soil is affected by physical and chemical characters of the soil, such as temperature, pH, moisture, organic matter and soil texture (Edwards and Bohlen, 1996).

2.3 Temperature

The activity, metabolism, growth, respiration and reproduction of earthworms are all influenced greatly by temperature (Edwards and Bohlen, 1996).

2.4 pH

pH is a vital factor that determines the distribution of earthworms as they are sensitive to the hydrogen ion concentration (Edwards and Bohlen, 1996; Chalasani *et al.*, 1998). pH and factors related to pH influence the distribution and abundance of earthworms in soil (Staaf, 1987). Several workers have stated that most species of earthworms prefer soils with a neutral pH (Jairajpuri, 1993; Edwards and Bohlen, 1996). There is a significant positive correlation between pH and the seasonal abundance of juveniles and young adults (Reddy and Pasha, 1993).

2.5 Moisture

Prevention of water loss is a major factor in earthworm survival as water constitutes 75-90% of the body weight of earthworms (Grant, 1955). However, they have considerable ability to survive adverse moisture conditions, either by moving to a region with more moisture (Valle *et al.*, 1997) or by means of aestivation (Baker *et al.*, 1992). Availability of soil moisture determines earthworm activity as earthworm species have different moisture requirements in different regions of the world. Soil moisture also influences the number and biomass of earthworms (Wood, 1974).

2.6 Organic matter

The distribution of earthworms is greatly influenced by the distribution of organic matter. Soils that are poor in organic matter do not usually support large numbers of earthworms (Edwards and Bohlen, 1996). Several workers have reported a strong positive correlation between earthworm number and biomass and the organic matter content of the soil (Doube *et al.*, 1997; Ismail, 2005).

2.7 Soil texture

Soil texture influences earthworm populations due to its effect on other properties, such as soil moisture relationships, nutrient status and cation exchange capacity, all of which have important influences on earthworm populations (Lavelle, 1992).

2.8 Effect of earthworms on soil quality

Earthworms, which improve soil productivity and fertility (Edwards *et al.*, 1995), have a critical influence on soil structure. Earthworms bring about physical, chemical and biological changes in the soil through their activities and thus are recognised as soil managers (Ismail, 2005).

2.9 Effects on physical properties of soil

Soil structure is greatly influenced by two major activities of earthworms:

1. Ingestion of soil, partial breakdown of organic matter, intimate mixing of these fractions and ejection of this material as surface or subsurface casts.
2. Burrowing through the soil and bringing subsoil to the surface.

During these processes, earthworms contribute to the formation of soil aggregates, improvement in soil aeration and porosity (Edwards and Bohlen, 1996). Earthworms contribute to soil aggregation mainly through the production of casts, although earthworm burrows can also contribute to aggregate stability since they are often lined with oriented clays and humic materials (Lachnicht and Hendrix; 2001). Most workers have agreed that earthworm casts contains more water-stable aggregates than the surrounding soil and by their activity influence both the drainage of water from soil and the moisture holding capacity of soil, both of which are important factors for plant productivity (Edwards and Bohlen, 1996; Lachnicht and Hendrix; 2001).

2.10 Effect on chemical properties of soil

Earthworms bring about mineralisation of organic matter and thereby release the nutrients in available forms that can be taken up by the plants (Edwards and Bohlen, 1996). Organic matter that passes through the earthworm gut is egested in their casts, which is broken down into much finer particles, so that a greater surface area of the organic matter is exposed to microbial decomposition (Martin, 1991). Earthworms have major influences on the nutrient cycling process in many ecosystems (Edwards and Bohlen, 1996). These are usually based on four scales (Lavelle and Martin, 1992):

1. during transit through the earthworm gut,
2. in freshly deposited earthworm casts,
3. in aging casts, and
4. during the long-term genesis of the whole soil profile.

Earthworms contribute nutrients in the form of nitrogenous wastes (Ismail, 2005). Their casts have higher base-exchangeable bases, phosphorus, exchangeable potassium and manganese and total exchangeable calcium. Earthworms favour nitrification since they increase bacterial population and soil aeration. The most important effect of earthworms may be the stimulation of microbial activity in casts that enhances the transformation of soluble nitrogen into microbial protein thereby preventing their loss through leaching to the lower horizons of the soil. C: N ratios of casts are lower than that of the surrounding soil (Bouché, 1983). Lee (1983) summarised the influence of earthworms on soil nitrogen and nitrogen cycling. According to him, nitrogenous products of earthworm metabolism are returned to the soil through casts, urine, mucoproteins and dead tissues of earthworms.

3. Earthworms and microorganisms

There is a complex inter-relationship between earthworms and microorganisms. Most of the species of microorganisms that occur in the alimentary canal of earthworms are the same as those in the soils in which the earthworms live. The microbial population in earthworm casts is greatly increased compared with the surrounding soil (Haynes, *et al.*, 1999). Earthworm casts usually have a greater population of fungi, actinomycetes and bacteria and higher enzyme activity than the surrounding soil (Lachnicht and Hendrix, 2001). Microbial activity in earthworm casts may have an important effect on soil crumb structure by increasing the stability of the worm-cast-soil relative to that of the surrounding soil (Edwards and Bohlen, 1996). Earthworms are very important in inoculating soils with microorganisms. Many microorganisms in the soil are in a dormant stage with low metabolic activity, awaiting suitable conditions like the earthworm gut (Lachnicht and Hendrix, 2001) or mucus (Lavelle *et al.*, 1983) to become active. Earthworms have been shown to increase the overall microbial respiration in soil, thereby enhancing microbial degradation of organic matter.

4. Earthworms and plant growth

Earthworms prepare the ground in an excellent manner for the growth of plants (Darwin, 1881). Darwin's findings that earthworms play a beneficial role in soil fertility that is important for plant growth have been acknowledged by many workers (Lee and Foster, 1991; Alban and Berry, 1994; Nooren *et al.*, 1995; Decaens *et al.*, 1999). Earthworms have beneficial effects on soil and many workers have attempted to demonstrate that these effects increase plant growth and yields of crops (Decaens *et al.*, 1999; Lalitha *et al.*, 2000;). Earthworms release substances beneficial to plant growth like auxins and cytokinins (Krishnamoorthy and Vajranabhaiah, 1986). The beneficial effect of earthworms on plant growth may be due to several reasons apart from the presence of macronutrients and micronutrients in vermicast and in their secretions in considerable quantities (Lalitha *et al.*, 2000; Ismail, 2005). Reports suggest that certain metabolites produced by earthworms may be responsible for stimulating plant growth.

5. Earthworms and land reclamation

The success of land reclamation by conventional techniques is often limited by poor soil structure and low inherent soil fertility, and even in productive soils, a marked deterioration in the botanical composition of the sward can occur within a number of years (Hoogerkamp *et al.*, 1983). A number of studies indicate that earthworms play an important part in improving reclaimed soils (Boyle *et al.*, 1997; Butt, 1999). Some experiments on improving impoverished soils by stimulating earthworm populations have been reported (Butt *et al.*, 1997). A successful introduction of earthworms in reclaimable soil could be achieved by overcoming factors like unfavorable moisture conditions, excessive fluctuation of surface temperature and lack of suitable food (Satchell, 1983).

6. Earthworms and organic solid waste management

In recent years, disposal of organic wastes from various sources like domestic, agriculture and industrial has caused serious environmental hazards and economic problems. Burning of organic wastes contributes tremendously to environmental pollution thus, leading to polluted air, water and land. This process also releases large amounts of carbon dioxide in the atmosphere, a main contributor to global warming together with dust particles. Burning also destroys the soil organic matter content, kills the microbial population and affects the physical properties of the soil (Livan and Thompson, 1997). It has been demonstrated that earthworms can process household garbage, city refuse, sewage sludge and waste from paper, wood and food industries (Kale *et al.*, 1982; Muyima *et al.*, 1994; Edwards and Bohlen, 1996; Ismail, 2005). In tropical and subtropical conditions *Eudrilus eugeniae* and *Perionyx excavatus* are the best vermicomposting earthworms for organic solid waste management (Kale, 1998). The use of earthworms in composting process decreases the time of stabilisation of the waste and produces an efficient bio-product, *i.e.,* vermicompost.

Organic farming system is gaining increased attention for its emphasis on food quality and soil health. Vermicompost and vermiculture associated with other biological inputs have been actually used to grow vegetables and other crops successfully and have been found to be economical and productive (Ismail, 2005; Ansari and Ismail, 2008).In this regard, recycling of organic waste is feasible to produce useful organic manure for agricultural application. Compost is becoming an important aspect in the quest to increase productivity of food in an environmentally friendly way. Compost is becoming an important aspect in the quest to increase productivity of food in an environmentally friendly way. Vermicomposting offers a solution to tonnes of organic agro-wastes that are being burned by farmers and to recycle and reuse these refuse to promote our agricultural development in more efficient, economical and environmentally friendly manner. Both the sugar and rice industries burn their wastes thereby, contributing tremendously to environmental pollution thus, leading to polluted air, water and land. This process also releases large amounts of carbon dioxide in the atmosphere, a main contributor to global warming together with dust particles. Burning also destroys the soil organic matter content, kills the microbial population and affects the physical properties of the soil (Livan and Thompson, 1997). Therefore organic farming helps to provide many advantages such as; eliminate the use of chemicals in the form of fertilizers/pesticides, recycle and regenerate waste into wealth; improve soil, plant, animal and human health; and creating an ecofriendly, sustainable and economical bio-system models (Ansari and Ismail, 2001a).

6.1 Vermitechnology

Vermitechnology is the use of surface and subsurface local varieties of earthworm in composting and management of soil (Ismail, 2005). Darwin (1881) has made their activities the object of a careful study and concluded that 'it may be doubted if there are any other animals which have played such an important part in the history of the world as these lowly organized creatures'. It has been recognized that the work of earthworms is of tremendous agricultural importance. Earthworms along with other animals have played an important role in regulating soil processes, maintaining soil fertility and in bringing about nutrient cycling (Ismail, 1997). Earthworms have a critical influence on soil structure, forming aggregates and improving the physical conditions for plant growth and nutrient uptake. They also improve soil fertility by accelerating decomposition of plant litter and organic matter and, consequently, releasing nutrients in the form that are available for uptake by plants.

6.2 Vermicomposting

Vermicomposting is the biological degradation and stabilization of organic waste by earthworms and microorganisms to form vermicompost. This is an essential part in organic farming today. It can be easily prepared, has excellent properties, and is harmless to plants. The earthworms fragment the organic waste substrates, stimulate microbial activity greatly and increase rates of mineralization. These rapidly convert the waste into humus-like substances with finer structure than thermophilic composts but possessing a greater and more diverse microbial activity. Vermicompost being a stable fine granular organic matter, when added to clay soil loosens the soil and improves the passage for the entry of air. The mucus associated with the cast being hydroscopic absorbs water and prevents water logging and improves water-holding capacity. The organic carbon in vermicompost releases the nutrients slowly and steadily into the system and enables the plant to absorb these nutrients. The soil enriched with vermincompost provides additional substances that are not found in chemical fertilizers (Kale, 1998). Vermicomposting offers a solution to tonnes of organic agro-wastes that are being burned by farmers and to recycle and reuse these refuse to promote our agricultural development in more efficient, economical and environmentally friendly manner. The role of earthworms in organic solid waste management has been well established since first highlighted by Darwin (1881) and the technology has been improvised to process the waste to produce an efficient bio-product vermicompost (Kale *et al.*, 1982; Ismail, 1993, Ismail, 2005). Epigeic earthworms like *Perionyx excavatus, Eisenia fetida, Lumbricus rubellus* and *Eudrilus eugeniae* are used for vermicomposting but the local species like *Perionyx excavatus* has proved efficient composting earthworms in tropical or sub-tropical conditions (Ismail, 1993; Kale, 1998). The method of vermicomposting involving a combination of local epigeic and anecic species of earthworms (*Perionyx excavatus* and *Lampito mauritii*) is called Vermitech (Ismail, 1993; Ismail, 2005). The compost prepared through the application of earthworms is called vermicompost and the technology of using local species of earthworms for culture or composting has been called Vermitech (Ismail, 1993). Vermicompost is usually a finely divided peat-like material with excellent structure, porosity, aeration, drainage and moisture holding capacity (Edwards, 1982, 1988). The nutrient content of vermicompost greatly depends on the input material. It usually contains higher levels of most of the mineral elements, which are in available forms than the parent material (Edwards and Bohlen, 1996). Vermicompost improves the physical, chemical and biological properties of soil (Kale, 1998). There is a good evidence that vermicompost promotes growth of plants (Lalitha *et al.*, 2000) and it has been found to have a favourable influence on all yield parameters of crops like wheat, paddy and sugarcane (Ismail, 2005).

Vermiculture is the culture of earthworms and vermicast is the fecal matter released by the earthworms (Ismail, 2005). Many agricultural industries use compost, cattle dung and other animal excreta to grow plants. In today's society, we are faced with the dilemma of getting rid of waste from our industries, household etc. In order for us to practice effective waste management we can utilize the technology of vermicomposting to effectively manage our waste. This process allows us to compost the degradable materials and at the same time utilize the products obtained after composting to enhance crop production and eliminate the use of chemical fertilizers. As indicated by Ansari and Ismail (2001), the application of chemical fertilizers over a period has resulted in poor soil health, reduction in produce, and increase in incidences of pest and disease and environmental pollution. In order to cope with these trenchant problems, the vermin-technology has become the most suitable remedial device (Edwards and Bohlen, 1996; Kumar, 2005).

6.3 Vermiwash

Vermiwash is a liquid that is collected after the passage of water through a column of worm action and is very useful as a foliar spray. It is a collection of excretory products and mucus secretion of earthworms along with micronutrients from the soil organic molecules. These are transported to the leaf, shoots and other parts of the plants in the natural ecosystem. Vermiwash, if collected properly, is a clear and transparent, pale yellow coloured fluid (Ismail, 1997). Vermiwash, a foliar spray, is a liquid fertilizer collected after the passage of water through a column of worm activation. It is a collection of excretory and secretory products of earthworms, along with major micronutrients of the soil and soil organic molecules that are useful for plants (Ismail, 1997). Vermiwash seems to possess an inherent property of acting not only as a fertilizer but also as a mild biocide (Pramoth, 1995).

7. Conclusion

Environmental Hazards are compounded by accumulation of organic waste from different sources like domestic, agricultural and industrial wastes that can be recycled by improvised and simple technologies. Vermicompost could be effectively used for the cultivation of many crops and vegetables, which could be a step towards sustainable organic farming. Such technologies in organic waste management would lead to zero waste techno farms without the organic waste being wasted and burned rather then would result in recycling and reutilization of precious organic waste bringing about bioconservation and biovitalization of natural resources.

8. References

Alban, D. H. and Berry, E. C. 1994. Effects of earthworm invasion on morphology, carbon and nitrogen of forest soil. *Appl. Soil Ecol.*, 1: 243- 249.

Ansari, A. A and S. A. Ismail. 2001a. Vermitechnology in Organic Solid Waste Management. Journal of Soil Biology and Ecology 21:21-24.

Ansari, A. A. and Ismail, S. A. 2001b. A Case Study on Organic Farming in Uttar Pradesh. Journal of Soil Biology and Ecology. 27: 25-27.

Ansari, A. A and Ismail, S. A. 2008. Reclamation of sodic soils through Vermitechnology. *Pakistan Journal of Agricultural Research*, Volume 21, Number (1-4): 92-97.

Ansari, A. A and Sukhraj, K. 2010. Effect of vermiwash and vermicompost on soil parameters and productivity of okra (*Abelmoschus esculentus*) in Guyana. *Pakistan Journal of Agricultural Research,* Volume 23 Number (3-4): 137-142.

Baker, G. H., Barret, V. J., Gray-Gardner, R. and Buckerfield, J.C. 1992. The life history and abundance of the introduced earthworms *Aporrectodea trapezoides* and *Aporrectodea caliginosa* in pasture soils in the Mount Lofty Range, South Australia. *Aust. J. Ecol.,* 17: 177-188.

* Bouché, M. B., 1971. Relations entre les structures spatiales et fonctionelles des écosystemes, illustrées par le rôle pédobiologique des vers de terre, In: *La Vie dans les Sols, Aspects Nouveaux, Études Experimentales,* (Pesson, P. ed.), Gauthier-Villars, Paris. pp: 187-209.

Bouché, M. B., 1977. *Strate'gies lombriciennes.* In: Soil organisms components of ecosystems. (Lohm, U., and Persson, T. eds.), *Biol Bull.,* Stockholm, 25: 122-132.

* Bouché, M. B., 1983. Ecophysiologie des lombriciens: Acquis récentes et perspectives In: *New trends in soil biology.* (Leprun, Ph., André, H. M., de Medts, A., Grégoire-Wibo, C. and Wauthy, G. eds.), pp: 321-333.

Boyle, K. E., Curry, J. P. and Farrell, E. P. 1997. Influence of earthworms on soil properties and gross production in reclaimed cutover peat. *Biol. and Fertil. Soils.,* 25: 20-26.

Butt, K. 1999. The effect of temperature on the intensive production of *Lumbricus terrestris* (Oligochaeta: Lumbricidae). *Pedobiologia.,* 35: 257-264.

Butt, K. R., Fredrickson, J., Morris, R. M. and Edwards, C. A. 1997. The earthworm inoculation unit technique: an integrated system for cultivation and soil-inoculation of earthworms. *Soil Biol. Biochem.,* 29: 251-257.

Chalasani, D., Krishna, S. R., Reddy, A. V. S. and Dutt. C. 1998. Vermiculture biotechnology for promoting sustainable agriculture. *Asia Pacific Journal of Rural Development.,* 8: 105-117.

Darwin, C., 1881. *The formation of vegetable mould through the action of worms, with observations on their habitats.* Murray, London. 326 pp.

Decaens, T., Jimenez, J. J., Lavelle, P., Diaz-Cosin, D. J., Jesus, J. B., Trigo, D. and Garvin, M. H. 1999. Effect of exclusion of the anecic earthworm *Martiodrilus carimaguensis* Jimenez and Moreno on soil properties and plant growth in grasslands of the eastern plains of Colombia. 6th International Symposium on Earthworm Ecology, Vigo, Spain, *Pedobiologia.,* 43 (6): 835-841.

Doube, B. M., Schimdt, O., Killham, K. and Correll, R. 1997. Influence of mineral soil on the palatability of organic matter for the lumbricid earhtworms: A simple food preference study. *Soil. Biol. Biochem.,* 29: 569-575.

Edwards, C. A. 1982. Production of earthworm protein for animal feed from potato waste. In: *Upgrading waste for feed and food.* (Ledward, D. A., Taylor, A. J. and Lawrie, R. A. eds.), Butterworths, London.

Edwards, C. A. 1988. Breakdown of animal, vegetable and industrial organic waste by earhtworms. *Agric. Ecosyst. Environ.,* 24: 21-31.

Edwards, C. A. and Bohlen, P. J. 1996. *Biology and ecology of earthworm.* (3rd edn.), Chapman and Hall, London. 426 pp.

Edwards, C. A. and Lofty, J. R. 1972. *Biology of earthworms.* Chapman & Hall, London. 283 pp.

Edwards, C. A., Bohlen, P. J., Linden, D. R. and Subler, S. 1995. Earthworms in agroecosystems. In: *Earthworm Ecology and Biogeography in North America.* (Hendrix, P. F. eds.), Lewis Publisher, Boca Raton, FL, pp: 185-213.

Grant, W. C. 1955. Studies on moisture relationships in earthworms. *Ecology.,* 36: 400-407.

Haynes, R. J., Fraser, P. M., Tregutha, R. J., Piercy, J. E., Diaz-Cosin, D. J., Jesus, J. B., Trigo, D. and Garvin, M. H. 1999. Size and the activity of the microbial biomass and N, S and P availability in earthworm casts derived from arable and pastoral soil amended with plant residues. 6th *International symposium on Earthworm Ecology., Vigo, Spain. Pedobiologia.*, 43: 568-573.

Ismail, S. A. 1993. Keynote Papers and Extended Abstracts. *Congress on traditional sciences and technologies of India*, I.I.T., Mumbai. 10: 27-30.

Ismail, S. A. 1997. *Vermicology: The Biology of Earthworms.* Orient longman Press, Hyderabad. 92 pp.

Ismail, S.A., 2005. The Earthworm Book. Other India Press, Mapusa, Goa. 101p.

Jairajpuri, M. S. 1993. Earthworms and vermiculture: an introduction. In: *Earthworm resources and vermiculture*, ZSI, Kolkata, India. pp: 1-5.

Kale, R. D. 1998. *Earthworm Cinderella of Organic Farming*. Prism Book Pvt Ltd, Bangalore, India. 88 pp.

Kale, R. D., Bano, K. and Krishnamoorthy, R. V. 1982. Potential of *Perionyx excavatus* for utilising organic wastes. *Pedobiologia.*, 23 : 419-425.

Krishnamoorthy, R. V. and Vajranabhaiah, S. N. 1986. Biological activity of earthworm casts: An assessment of plant growth promoter levels in the casts. *Proc. Indian Acad. Sci.(Anim. Sci.).*, 95: 341-351.

Kumar, A. 2005. Verms and Vermitechnology. Vedams eBooks (P) Ltd, New Delhi India. pp. 110-034

Lachnicht, S. L. and Hendrix, P. F. 2001. Interaction of earthworm *Diplocardia mississippiensis* (Megascolecidae) with microbial and nutrient dynamics in subtropical Spodosol. *Soil Biol. Biochem.*, 33: 1411-1417.

Lalitha, R., Fathima, K. and Ismail, S. A. 2000. Impact of biopesticides and microbial fertilizers on productivity and growth of *Abelmoschus esculentus. Vasundhara The Earth.*, 1 & 2: 4-9.

Lavelle, P. 1992. *Conservation of soil fertility in low-input agricultural systems of the humid tropics by manipulating earthworm communities (macrofauna project).* European Economic Community Project No.TS2-0292-F (EDB).

Lavelle, P. and Martin, A. 1992. Small scale and large scale effects of endogeic earthworms on soil organic matter dynamics in soil of the humid tropics. *Soil Biol. Biochem.*, 24: 1491-1498.

Lavelle, P., Rangel, P. and Kanyonyo, J. 1983. Intestinal mucus production by two species of tropical earthworms: M. lamtoniana and P. corethrurus. In: *New Trends in Soil Biology*, (Lebrun, *P.* eds.), Dieu-Brichart Press, Louvain le Neuve, Belgium, pp: 405-410.

Lee, K. E. 1983. The influence of earthworms and termites on soil nitrogen cycling. In: *New trends in Soil Biology.* (Lebrun, P. H., Andre, H. M., de Medts, A., Gregoire-Wibo, C. and Wathy, G. eds.), pp: 35-48.

Lee, K. E. 1985. *Earthworms: Their ecology and relationships with soils and land use.* Academic Press, Sydney. 411 pp.

Lee, K. and Foster, R. C. 1991. Soil fauna and soil structure. *Aust. J. Soil Res.*, 29: 745-776.

Livan, M. A and W. Thompson. 1997. NARI Annual Report.

Martin, A. 1991. Short- and long-term effects of the endogeic earthworm *Millsonia anomala* (Omodeo) (Megascolecidae, Oligochaeta) of tropical savannas on soil organic matter. *Biol. Fertil. Soils.*, 11: 234-238.

Muyima, N. Y. O., Reinecke, A. J. and Viljoen-Reinecke, S. A. 1994. Moisture requirements of *Dendrobaena veneta*- a candidate for vermicomposting. *Soil Biol. Biochem.*, 26: 973-976.

Nooren, C. A. M., Van Breeman, N., Stoorvogel, J. J. and Jongmans, A. G. 1995. The role of earthworms in the formation of sandy surface soils in a tropical forest in Ivory Coast. *Geoderma.*, 65: 135-148.

Piearce, T. G., Oates, K. and Carruthers, W. J. 1990. A fossil earthworm embryo (Oligochaeta) from beneath a late bronze age midden at Potterna, Wiltshire, UK. *J. Zool. Land.*, 220: 537-542.

Pramoth, A. 1995. *Vermiwash-A potent bio-organic liquid "Ferticide".* M.Sc., dissertation, University of Madras. 29 pp.

Reddy, M. V. and Pasha, M. 1993. Influence of rainfall, temperature and some soil physicochemical variables on seasonal population structure and vertical distribution of earthworms in two semi- and tropical grassland soils. *Int. J. Biotech.*, 37: 19-26.

Satchell, J. E. 1983. Earthworm microbiology. In: *Earthworm ecology: From Darwin to Vermiculture.* (Satchell, J. E. ed.), Chapman and Hall, London, UK. pp: 351-364.

Shipley, A. E. 1970. In: *The Cambridge Natural History.* (Harmer, S. F. and Shipley, A. E. eds.). Codicote, England.

Shuster, W. D., Subler, S. and McCoy, E. L. 2000. Foraging by deep-burrowing earthworms degrades surface soil structure of a fluventic Hapludoll in Ohio. *Soil & Tillage Research.*, 54: 179-189.

Staaf, H. 1987. Foliage litter turnover and earthworm populations in three beech forests of contrasting soil and vegetation types. *Oecologia.*, 72: 58-64.

Tomlin, A. D., Shipitalo, M. J., Edwards, W. M. and Protz, R. 1995. Earthworms and their influence on soil structure and infilteration. In: *Earthworm ecology and biogeography in North America.* (Hendrix, P.F. ed.), Lewis Publishers, Chelsea. pp: 159-184.

Valle, J. V., Moro, R. P., Garvin, M. H., Trigo, D. and Diaz Cosin, D. J. 1997. Annual dynamics of the earthworms *Hormogaster elisae* (Oligochaeta, Hormogastridae) in Central Spain. *Soil Biol. Biochem.*, 29: 309-312.

Wood, T. G. 1974. The distribution of earthworms (Megascolecidae) in relation to soils, vegetation and altitude on the slopes of Mt. Kosciusko, Australia. *J. Anim. Ecol.*, 43: 87-106.

* Not referred directly

6

Co-Digestion of Organic Waste and Sewage Sludge by Dry Batch Anaerobic Treatment

Beatrix Rózsáné Szűcs[1], Miklós Simon[1]
and György Füleky[2]
[1]*Eötvös József College*
[2]*Szent István University*
Hungary

1. Introduction

Organic waste and waste water sludges can be stabilized both in anaerobic and in aerobic processes. The advantage of the anaerobic treatment of the waste is that biogas develops during the degradation process. Instead of energy consumption which is a usual characteristic of aerobic-processes it is accompanied with energy production, which can be utilized as energy source (Kayhanian & Tchobanoglous, 1992; Cout et al., 1994).

The anaerobic processes can be further classified according to the dry mater content and the feeding of the fermenting reactor. According to Tchobanoglous (1993) and his work team, we can talk about semidry procedures in the range of 15-20% dry matter contents. If the dry matter content is high, only batch reactors operated with the principle of filling and emptying can be applied, in order to avoid difficulties due to the continuous feeding.

On waste water treatment plants of small and medium capacity, the waste water sludge can not be economically stabilized by the conventional anaerobic treatment of low dry matter content and continuous feeding. Thus, the sludge is usually stabilized by composting in that cases. At waste water treatment plants of great capacity, the sludge is stabilized by anaerobic treatment of liquid, continuous technology which is often followed by composting, in order to achieve better material characteristics of the end product.

The municipal waste management directives require that the organic content of the wastes to be dumped should be reduced. The realization of the waste management goals requires the stabilization of municipal organic wastes, where generally composting is applied. The sewage sludge and the organic fraction of municipal solid waste, called vegetable, fruit and garden (VFG) waste are different from each other regarding their materials and quality, yet for their stabilization, combined treatment is more and more often applied. The quantity and quality of the VFG varies with time and space, depending on the season, the structure of the settlement and the standard of living.

The novel dry batch BIOCEL technology was introduced for the treatment of municipal solid organic waste in the Netherlands. It has the advantage that it is simple to operate, and its specific reactor volume projected to the treated material flow is low (Brummeler et al., 1991; Brummeler, 2000; Simon, 2000). The investment costs of the dry batch BIOCEL technology are lower by 40% than those of the continuous anaerobic systems (Brummeler et al., 1992). Its advantage over composting by state-of-the-art technology comes from a simpler technical solution and a more economical operation.

We assume that the dry batch anaerobic treatment could be used for combined anaerobic treatment of VFG wastes and municipal sewage sludge. When treating the waste water sludge and other municipal organic wastes together by anaerobic method, the possible too high easily degradable organic content of the VFG might be a problem, because in lack of sufficient seeding material, it can lead to acidification of the system. A number of literature reports about the anaerobic treatment of different organic wastes separately (Brummeler, 1993), but there are no results available regarding co-digestion by dry anaerobic treatment.

The effective anaerobic conversion of organic substances into methane depends on the activity of miscellaneous microbial populations. A diagram of the consecutive metabolic stages, which can be distinguished in anaerobic digestion, is shown in figure 1. In well balanced digestion, all products of a previous metabolic stage are converted into the next one. The overall result is a nearly complete conversion of the biodegradable organic material in the waste into end products such as methane, carbon dioxide, hydrogen sulphide, ammonia, etc. without significant build-up of intermediate products.

The products of the fermentation vary depending on quality of raw material and environmental conditions applied. Low pH values decreases the relative amount of acetic acid and increases the relative amount of propionic acid (Breure & van Andel, 1984). Partial pressure of hydrogen in the gas phase can significantly influence the kind of products formed by fermentative bacteria (Wolin & Miller, 1982). Hanaki et al. (1981) stated that β-oxidation of long-chain fatty acids is thermodynamically unfavourable unless the hydrogen partial pressure is maintained at a very low level. The dependence of a low hydrogen partial pressure makes, that long chain fatty acids degradation can be inhibited indirectly by inhibition of hydrogen consuming organisms (Koster, 1989).

The high substrate affinity of the hydrogen-consuming micro organisms makes it possible to maintain low hydrogen concentrations. According to Robinson & Tiedje (1982), the Michaelis-Menten half-saturation constant (K_m) for hydrogen is in the range of 5.8-7.1 μM. Zehnder et al. (1982) stated, that in a well balanced methane fermentation, the hydrogen partial pressure does not exceed 10^{-4} atm and in most cases approximately 10^{-6} atm.

Macro and micro nutrients contents of the treated raw material effects the yield of micro organisms. Lettinga et al. (1981) advised for macro nutrients a COD to nitrogen to phosphorus ratio in a range of 350 : 5 : 1 – 1000 : 5 : 1.

The precondition of the efficient application of the anaerobic batch reactors is the establishment of the balance between the acid production and the methane production, in the absence of the reactor getting acidified (Benedek, 1990). During the anaerobic degradation of the organic material, four consecutive metabolic steps can be distinguished: hydrolysis, acidogenesis, acetogenesis, and methanogenesis (Batstone et al., 2002).

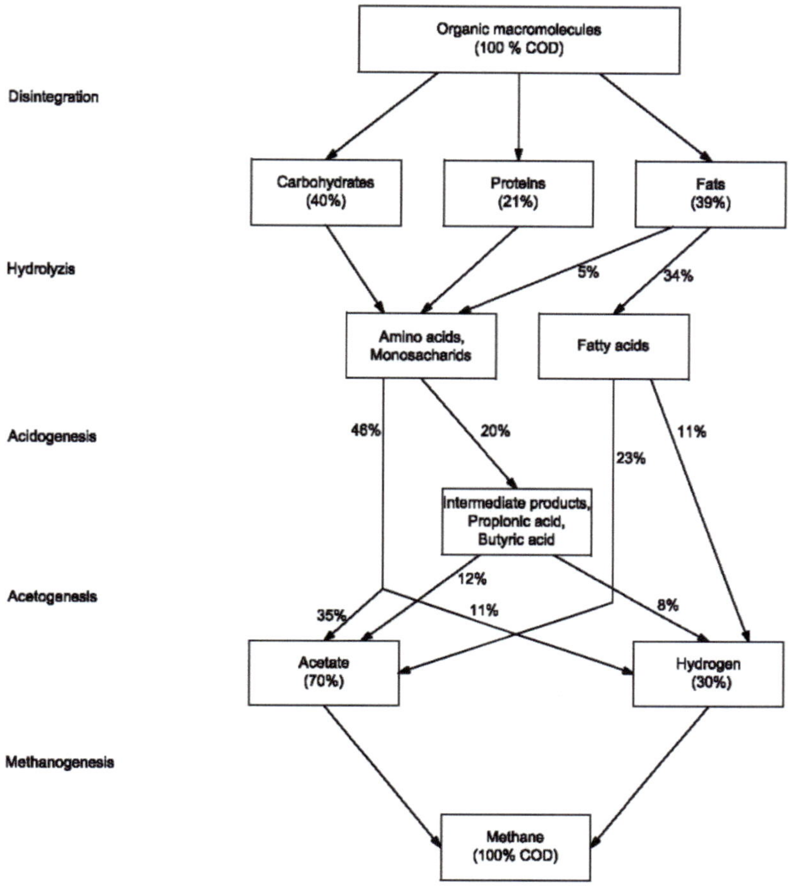

Fig. 1. Metabolic stages and products in anaerobic digestion of complex organic material

Among them, the usual rate-limiting factor of the whole procedure is the methagonesis (Gosh & Klass, 1978). The rate of the acid production is great compared to that of the methane production. At a balanced anaerobic degradation process, the elimination rate of the biologically degradable organic dry material is almost equal to that of the methane production (Gujer & Zehnder, 1983), because the biomass production is negligible.

The research and the thermodynamic calculations show that 70% of the methane is generated during the decarboxylation of the acetic acid, and the remaining 30% comes from the reduction of the carbon dioxide (Jeris & McCarthy, 1965; Kaspar & Wuhrmann, 1978).

$$CH_3COOH \rightarrow CH_4 + CO_2 \qquad (\Delta G° = -39{,}5\ kJ)$$

$$CO_2 + 4\ H_2 \rightarrow CH_4 + 2\ H_2O \qquad (\Delta G° = -145\ kJ)$$

So, the efficiency of the methane production is characterized not only by measuring the methane content, but also by determining the hydrogen content. In case the hydrogen is accumulated and is not converted to methane, then the accumulated hydrogen will immediately inhibit the oxidation of the propionic acid and the accumulated acetic acid. This will result in decrease in pH and, thus, in acidification of the reactors. This again will affect the oxidation of the hydrogen, decreasing the reaction efficiency and increasing the partial pressure of the hydrogen (Gujer & Zehnder, 1983).

To keep the degradation process balanced, seeding material is needed. A significant effect in balancing the process can be achieved by properly setting the ratio of the methanogen seeding material. However, the determination of the optimal amount of the seeding material is a complex issue which is of great importance to the operation of batch reactors being economical. Low seeding material ratio, in extreme cases, can lead to acidification of the reactor or, in better cases, moderates the process rate. This can be compensated by extending the retention time associated with increased reactor volume.

The increase of the biogas production and the decrease of the treatment time can be achieved also by increasing the quantity of the seeding material since this way a more effective degradation can be counted on. However, increasing the quantity of the seeding material can result in the increase of the reactor volume, too.

The optimal waste to seeding material ratio in the case of municipal solid organic wastes is 1:2.3 in laboratory, while less than 1 : 1 in full scale conditions (Brummeler, 1993). According to related literature data, the duration of the treatment in cases of municipal organic waste is around 30-36 days (Brummeler et al., 1991, 1992); however in cases of a low seeding material ratio, the duration of the treatment can be 50 days or more (Brummeler et al., 1992). There is no published data about seeding material demand for the dry, batch anaerobic co-treatment of the biowaste and waste water sludge.

We assume that the combined dry batch treatment of VFG waste generated on settlements and of sewage sludge has many advantages. As a result of the co-digestion, because of the different easily degradable organic contents of the sewage sludge and VFG, we can count on the increase of the gas yield projected to reactor volume, compared to a separate treatment of the VFG and sludge. We can assume as an advantage that a more balanced quality of the sewage sludge can have a positive effect on the co-digestion with organic wastes having quality varying with time and space. A further advantage can be, from the aspect as a potential of anaerobic treatment of the sewage sludges generated on smaller settlements, that the increased waste flow with VFG can make it economical. Nevertheless, it is necessary to investigate the appropriate seeding material ratio, the determination of which does not depend only on the achievable methane yield but on the required duration of the treatment and on the targeted stabilization goal of organic material, too.

Our aim is to study the combined dry batch treatment of VFG and sewage sludge. Our goal is to evaluate the aspects of determination of the optimal seeding material ratio, besides the study of the avoidance of acidification of the reactors, the achievable greater degradation rate of organic material and the maximal gas yield.

2. Materials and methods

In order to achieve our goals, we carried out laboratory experiments with dry batch reactors.

2.1 Materials

To ensure the repeatability of experiments, we modelled the biowaste (mixture of sewage sludge and VFG) generated in the settlements with a material mixture of fixed ratio as follows: 50% municipal excess sludge, 50% VFG consisting 25% fresh grass and 25% kitchen waste. The excess sludge came from the activated sludge technology of a municipal waste water plant, which can be characterized with a 20-day sludge retention time. The kitchen waste consisted of 25% potato peel, 15% lettuce, 15% bread, 15% cucumber peel, 10% cabbage, 10% paper and 10% coffee grounds.

The amount of total solids (furthermore as TS), volatile solids (furthermore as VS) and the value of chemical oxygen demand (furthermore as COD) of the waste and the sludge are presented in Table 1.

Materials	Total solids (TS %)	Volatile solids (VS %)	Chemical oxygen demand (COD) ($g\ O_2 \cdot kg\ TS^{-1}$)
methanogenic seed (digested sludge)	24.54	56.86	667
excess sludge	28.87	48.22	556
fresh grass	31.30	92.20	985
potato peel	18.51	94.33	1 074
lettuce	7.69	85.67	1 193
bread	65.35	97.36	1 094
cucumber peel	4.82	84.44	1 486
cabbage	8.79	90.98	1 086
paper	92.49	98.98	1 288
coffee grounds	34.30	99.28	1 145

Table 1. Characteristics of waste and sludge used for the experiment

To characterize the seed, we defined its stability and methanogenic activity. The seed was not stable, it could be degraded by a further 13%. The organic degradation occurred mostly within the first 30 days. The digested sludge came from a completely stirred tank reactor operated with 20 days hydraulic retention time. The methanogenic activity of the seed was 0.026 CH_4-COD $\cdot$ g $VS^{-1} \cdot d^{-1}$, which shows the maximum methane production measured in chemical oxygen demand (COD) of digested sludge for a unit of volatile solid in a unit of time.

2.2 Methods

The TS content and the volatile solids (VS) content of the samples were determined by drying and burning to constant weight at a temperature of 105°C and 650°C, respectively. The chemical oxygen demand (COD) of the sludges was measured by the standard method MSZ 21976-10:1982.

The amount of biogas generated, was measured by an „A1" type, Schlumberger wet gas meter. The methane and hydrogen content of biogas was measured by a Shimadzu 2014 gas chromatograph. The temperature of the column was 60°C, the temperature of the injector was 170°C, and the temperature of the detector was 250°C. As carrier gas we used nitrogen with 20 mL/min gas flow. In the 3.0-m long, 3.00-mm internal diameter glass column, Supelco Molecular sieve filling was put. The detection was done with TCD detector. We measured the quantity and the methane content of the biogas every day at the beginning, and then, when the amount of the biogas decreased, every other and then every fifth day.

To determine the methanogenic activity of the seeding material we used neutralized acetic acid as a substrate. To decrease the retardatory effects we added macro- and micro-nutrients (Biotechnion, 1996), and incubated the samples on the temperature of 35°C. We used liquid-phased mixed reactors to decrease the substrate-gradient. The amount of biogas generated, was calculated on the basis of pressure changes in the head-part of the 1.5 dm³ reactors. To remove the generated CO_2, NaOH pellets were placed in the heads-part of the reactors. Specific methanogenic activity of the seed was calculated on the basis of cumulative methane production graphs by taking the tangent of the deepest slope of the curve.

The acidity of the sludge was checked by a pH meter (340i WTW) pH/mV measuring device, to which a SenTix 41 type electrode was connected.

2.3 The experimental setup

We performed the examination of the effects of the seeding material on the dry batch anaerobic treatment by a series of reactors of a total capacity of 6 dm³, which consisted of 4 reactors, each of a capacity of 1.5 dm³, connected in parallel. By these set ups, the disturbing effects (opening of reactors) occurring during the pH measurements were reduced (Figure 2). The reactors were connected to gas-collecting bags.

Fig. 2. Set of dry batch anaerobic reactors

As experimental variable we checked five different seeding material ratios. We set the organic waste to seeding material ratios projected to the quantity of dry organic material, to these values: 1:0.5; 1:1; 1:1.5; 1:2, and 1:3. We measured the gas production of the seeding material (digested sludge) in a control reactor, thus, the degradation rate of the biowaste could be calculated separately. We compared the treatments when the sewage sludge is treated alone and when is co-digested with VFG waste, with the 1:1 seeding material ratio usually applied in the practice for the anaerobic treatment of municipal organic wastes. We kept the reactors in a room of a constant temperature of 34°C.

Each reactor was filled with an equal amount and quality (TS=22%) of waste. In order to prevent the disturbing effects caused by the oxygen, we flushed the heads of the reactors with nitrogen gas after the sampling. The diluting effect of the head-space was considered at the calculation of the results.

2.4 The quantification of anaerobic degradation

The COD of methane produced in the anaerobic degradation of organic substrate corresponds with the COD of the removed organic mass (Lettinga & Hulshoff Pol, 1990). The amount of organic matter removed during the anaerobic treatment, the degree of degradation, was determined by measuring the total amount of methane produced during the period (T), which was converted to COD, taking into account that 1 Ndm^3 methane is equivalent to 2.86 g CH_4-COD. Based on this, the degree of degradation of the organic material was defined by the formula below:

$$D_T\% = (\sum CH_4\ COD_T\ /\ \text{sludge COD}) \times 100 \quad (1)$$

We fitted a logistic function-relation ($D_T\% = D_{max} / (1 + e^{-k(t-t0)})$) to the measuring results with SPSS 14.0 software. We used sludge as seeding material and the substrate for the tests after storing at 5°C, therefore we had to calculate with the lag phase in the beginning by choosing the logistic function-relation. The logistic curves take into consideration the start-up phase, pursuant to the Monod and the Briggs-Haldane model.

We determined the value of maximum degradation (D_{max}) in case of biowaste and sludge for the fitting as 65% and 50% respectively, which values were based on our former own measuring results (Rózsané et al, 2011) and on technical literature data (Haug, 1980). We determined the k invariant of reaction speed and the t_0 time defining the inflexion point in a way that the function-relation would have the best fit (R^2) of the measuring results.

In the case of methane production projected to the volume of reactor, we did not deduce the methane production of the seeding material, but we used the results for the whole volume of the mixture of waste and seeding sludge. In case of the measuring results used for the volume of the reactor, we fitted the function-relation in a way identical with the previous, where the maximal degree of methane production (CH_{4max}) was determined with the account of substrate to seeding material mixing rates and the maximal degradability. To characterize the speed of the degradation process, in the case of both measuring results, we determined the values of the starting v_{10d} and v_{30d} degradation speed as the direction tangent to the fitted curves.

3. Results and evaluation

We assumed that the balance of the multi-stage anaerobic digestion process can be influenced by setting the ratio of the seeding material which results in greater degradation of the organic content of the treated waste, as well as in greater methane production.

We evaluated the experimental results based on two aspects:

- based on the degradation of the organic material achievable with different seeding material ratios; and

- based on the gas production achievable during the treatment, referred to the unit of reactor volume.

3.1 The results of the organic matter degradation

The actual methane production of different mixtures of organic wastes and seed, referred to one unit of treated organic material, is shown in Figure 3. The methane production of the seeding material present in the reactor was deducted from the methane production of the mixture of the waste and seeding material. As a result, because of the relatively high degradability of the seeding material, in the case of unbalanced reactors caused by low seeding material ratios, we had even negative methane production in the first 20 days which was indicated as zero value.

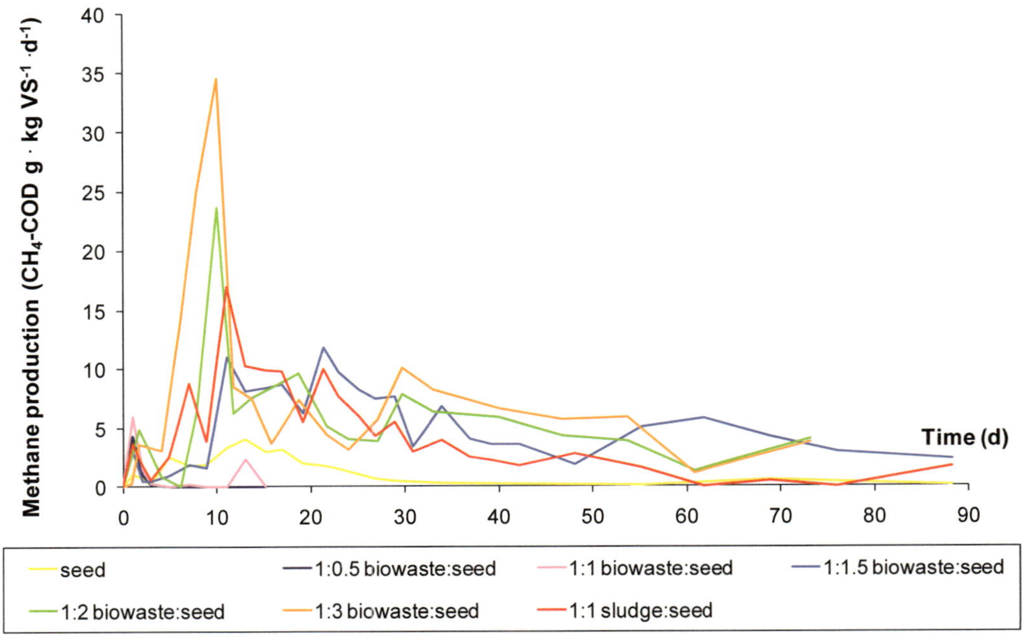

Fig. 3. Actual methane production referred to one unit of treated organic material

We reached the highest methane yield with the 1:3 biowaste to seed ratio. With the increase of the seed ratio, the methane production grew, too. The methane yield was very low in the case of 1:0.5 and 1:1 biowaste to seed ratios. Due to the low seed ratio, the waste became acidified (pH 5.5-5.8), so thus the process of methane production was also inhibited. Since our goal was to determine the optimal seeding ratio, we carried out the test in these reactors only for 15 days. The maximal methane production of the seeding material (digested sludge) occurred on day 10, however its extent was one eighth of that of the balanced reactors and the methane production decreased to zero after the 30 days.

Having compared the treatability of the sewage sludge and of the biowaste, with the 1:1 seed ratio applied in practice, we can state that in the case of the sewage sludge, a more balanced reactor performance can be observed. The results suggest that in the case of reactor containing VFG as well, the easily degradable organic material content was higher than in the case of the reactor containing only sewage sludge. The fatty acid accumulated the in reactor containing VFG which led to the acidification of the reactor, in the end. Against the acidification of the biowaste, in the case of the sludge, the values of pH and hydrogen concentration were better than the critical level even in the initial critical phase of the treatment. This calls the attention to that, because of the varying quality of VFG waste, the determination of the seeding material ratio has to be estimated case by case in each practical application.

We can calculate the degradation of organic material of the waste from the quotient of the methane production totalled in the time and of the chemical oxygen demand of the waste mixture. Figure 4 shows the rate of degradation against time for different seeding rates and substrates (the methane production of the seeding material is deducted). Onto the measurement results we fitted the logistic function describing biological processes ($Dt\% = D_{max} / (1 + e^{-k(t-t0)})$). The reaction kinetic parameters are shown in Table 2. The value of k reaction rate constant rose with the increase of the seeding material ratio which resulted in decrease of the value of t_0. Significant differences cannot be detected in the values of k and t_0 of the 1:1.5 and 1:2 mixing ratios.

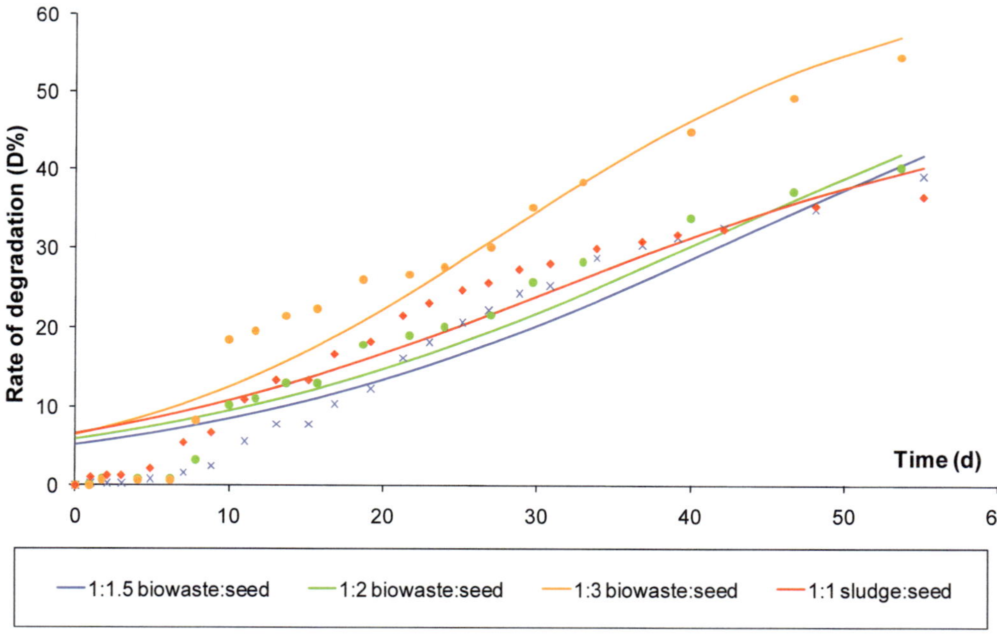

Fig. 4. The degradation rate of organic wastes against time

According to our measurement results, with a 60-days treatment with 1:3 biowaste to seed ratio, 54% organic material degradation can be achieved. In the case of biowaste to seed mixtures of 1:1.5 and 1:2 ratios, only 41-43% of the organic material became decomposed during the same period of time. Thus, when increasing the amount of seeding material with improving the initial conditions of the treatment, a considerable impact in the degradation rate of organic material can be achieved for the whole treatment period.

The description of the sample	k ($1 \cdot d^{-1}$)	t_0 (d)	R^2	V_{10d} ($D\% \cdot d^{-1}$)	V_{30d} ($D\% \cdot d^{-1}$)
1:1.5 biowaste:seed	0.055	44.33	0.936	0.424	0.788
1:2 biowaste:seed	0.054	42.58	0.918	0.453	0.782
1:3 biowaste:seed	0.078	28.29	0.939	0.820	1.256
1:1 sludge:seed	0.060	31.38	0.876	0.526	0.749

Table 2. Kinetic parameters of the degradation process

Considering the rates of actual methane production, we can see that the actual rates measured on day 10 significantly increase with the growth of the amount of seeding material. At the values related to day 30, the effect of seed ratio onto the methane production can be still well detected. The actual rate of the methane production further increased from day 30 also in each cases, which suggests that we can count on a considerable degree of degradation even after day 30. This is confirmed by the t_0 value, as well.

To characterize the process of the anaerobic degradation, we checked the hydrogen content of the biogas, as well as the temporal evolution of the pH of the reactors in the most critical initial phase of the treatment (Table 3). The hydrogen content of the biogas was above the value of the detection limit only in the first 9 days.

Type of the reactor	2nd day		3rd day		5th day		7th day		9th day	
	H_2 (%)	pH	H_2 (%)	pH	H_2 (%)	pH	H_2 (%)	pH	H_2 (%)	pH
1:0.5 biowaste:seed	9.66	5.47	0.38	5.51	1.08	5.55	0.08	5.65	0.06	5.70
1:1 biowaste:seed	4.27	5.70	0.19	5.80	0.13	5.75	0.02	5.78	0.02	5.83
1:1.5 biowaste:seed	3.58	5.90	0.14	5.84	0.10	6.13	0.02	6.06	0.01	6.34
1:2 biowaste:seed	1.40	6.27	0.04	6.32	0.02	6.34	0.01	6.43	0.01	6.59
1:3 biowaste:seed	0.37	6.10	0.05	6.23	<dl	6.45	0.01	6.65	0.01	6.72
1:1 sludge:seed	0.62	6.68	<dl	6.94	<dl	7.09	<dl	7.28	<dl	7.37

Table 3. The hydrogen content of biogas and the pH of wastes in the case of different wastes and seeding ratios

It is seen in the case of biowaste that, by the increase of seeding ratio, the hydrogen content of the biogas decreases and the pH of the waste in reactors increased. During the test period, the hydrogen content of the biogas also decreases and then, following day 9, it is under the value of detection limit. The critical hydrogen concentration, above the 0.01 % as calculated based on the literature (Zehnder et al., 1982), measured in the first 5 days had a negative effect on the methane production in the case of each seeding ratio (Figure 3). By the increase of the seeding, above the ratio of 1:1.5, the hydrogen concentration decreased below the critical value from day 9 and the methane production started to increase. The unfavourable values of hydrogen and pH measured in the case of 1:0.5 and 1:1 ratios led to the acidification of the reactors. The pH of the reactors increased during the test which resulted in the rise of biogas production. During the anaerobic treatment of the sewage sludge, we did not measure significant hydrogen quantity in the biogas even in the case of 1:1 sludge to seed mixing ratio. This can be explained by that there is less easily degradable organic material in the sewage sludge than in the tested biowaste which is responsible for the accumulation of hydrogen and volatile fatty acids. At sewage sludge digestion, often the hydrolysis appears as the process limiting step (Koster, 1989) which could contribute to the more favourable values measured in the case of sludge.

We compared our measurement results with the operation data of a full scale BIOCEL plant (Brummeler, 1993) (Table 4.). The literature refers the biogas quantities to wet waste mass, to standard condition. For comparability we recalculated the literature data to the CH_4-COD g · kg VS^{-1} unit used by us.

Time (d)	Cumulative methane production (CH_4-COD g · kg VS^{-1})			Degradation of the organic material (D%)		
	Full scale BIOCEL plant[1]	Laboratory scale		Full scale BIOCEL plant[2]	Laboratory scale	
		1:3 seed to biowaste[2]	1:1 seed to sludge		1:3 seed to biowaste[2]	1:1 seed to sludge
5	110.0	22.9	18.4	7.7	1.2	1.6
10	297.7	157.2	98.1	20.9	13.4	8.4
20	660.0	278.7	173.9	46.3	23.7	14.9
40	-	453.5	286.5	-	38.6	24.6
60	-	551.6	325.9	-	47.0	28.0

[1]Value calculated according to Brummeler (1993), 450 m^3 reactor, waste TS 36%, VS 65%

[2]Methane production together with the methane production of the seeding material

Table 4. Comparison of the laboratory results with the operation data of a full scale BIOCEL plant treating VFG

The results of Table 4 show that the results of the methane production referring to VFG waste reviewed in the literature, at the same moment, significantly exceed the results of the co-digestion of the biowaste with sewage sludge. In our experiment, the difference resulting from the lag phase as well as the lower degradability of the biowaste containing sludge and VFG can be definitely pointed. In our experiment half of the waste mixture was sewage sludge. The sewage sludge applied by us was less degradable than the biowaste, thus, the degradability of one unit of waste mixture (and so the amount of methane production from it, too) was lower.

The results of Table 4 show that higher gas yield referred to one unit of organic matter can be reached in the case of co-digestion of VFG and sewage sludge than in the case of sewage sludge digestion alone.

3.2 The results of methane production referred to reactor volume

We assume that the seed ratio, as a result of two opposing effects, influences the methane production per reactor volume unit. The increase of the seed ratio makes the anaerobic process balanced but at the same time decreases the amount of degradable organic matter per reactor volume unit. That is a question, to what extent the already digested material should be recycled for seeding. Another question is how the co-digestion of the easily degradable VFG and sewage sludge affects the gas production of the reactors. To answer the question, we checked the values of the totalled methane production referred to reactor volume unit for the sewage sludge and biowaste (consisting of 50% sludge and 50% VFG) at seed ratios shown in Figure 5.

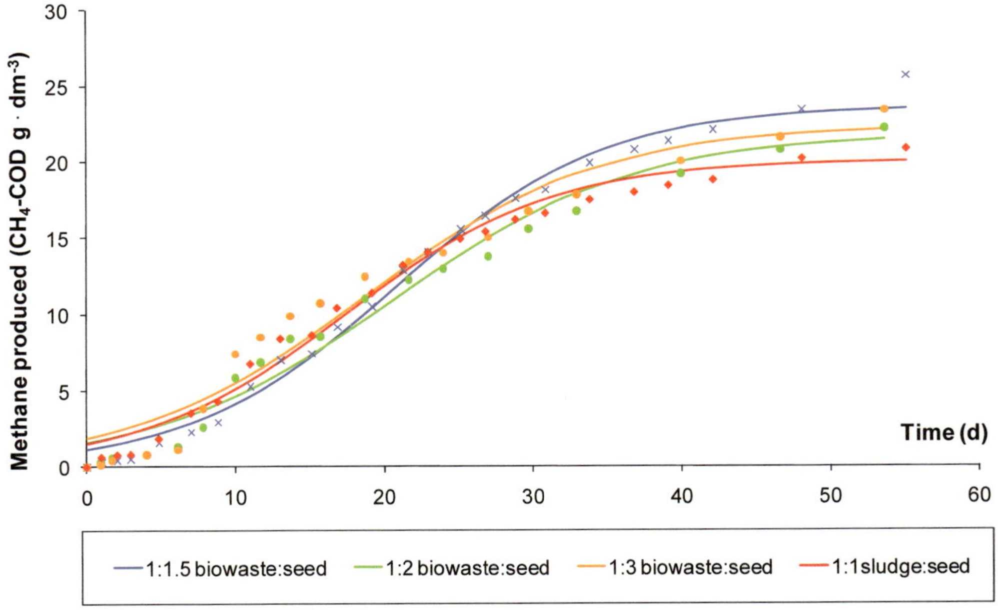

Fig. 5. Summarized, specific methane production on the basis of reactor volume against time

Figure 5 shows that in the case of methane production referred to reactor volume, the methane production of the sewage sludge above 30-day retention time is lower than that of the combined treatments, so thus, we can achieve higher gas yield from one unit of reactor volume when the sewage sludge and the biowastes are treated together than in case the sewage sludge is applied alone. We achieved maximal methane production with co-digestion at the 1:1.5 waste to seed ratio, this is followed by the 1:3 and 1:2 waste to seed ratios, however, significant difference between the measurement results cannot be detected. The increase of the seed ratio, in spite of the more inert material filling up the reactor volume, did not considerably reduce the methane production projected to reactor volume unit until day 30 of the treatment. A great increase of the amount of the seeding material, however, results in increase of the reactor volume necessary to the actual treatment capacity which, at the same time, is associated with the same rate of increase in gas production. Taking into consideration also the goal of stabilization, based on the comparison of Figures 4 and 5, we can state that is may be worth to count on the reduction of the retention time while increasing the seed ratio, for the purpose of optimization of the gas yield, degradation and volume demand.

The reaction kinetic parameters of results referred to the reactor volume are shown in Table 5. It is apparent from the results of the table that the values of the maximal methane production are nearly the same, no significant differences can be detected. The value of k reaction rate constant is the highest in the case of 1:1.5 biowaste to seed ratio and its value equals to the k value relating to the sewage sludge.

Type of the reactor	CH_4 produced (CH_4-COD g · dm^{-3})	v_{10d} (CH_4-COD g · dm^{-3} · d^{-1})	v_{30d} (CH_4-COD g · dm^{-3} · d^{-1})	k (1 · d^{-1})	t_0 (d)
1:1.5 biowaste:seed	23.56	0.529	0.503	0.143	20.85
1:2 biowaste:seed	21.68	0.505	0.485	0.123	20.59
1:3 biowaste:seed	22.26	0.586	0.442	0.127	18.75
1:1 sludge:seed	19.97	0.578	0.305	0.144	17.49

Table 5. Kinetic parameters of summarized methane production appertaining to the volume of the reactor

In the case of the values of actual methane production relating to day 10, we did not gain in each case higher v_{10} value when increasing the seeding ratio. The v_{30} value relating to day 30 is in all cases less than the v_{10} value which indicates the decrease of methane production. In the case of v_{30} values, we experienced that, when increasing the seeding, the value of actual methane production referred to reactor volume unit and relating to day 30 decreased.

We assume that the treatment period (retention time) affects the gas production of the reactor. The question is, taking into account the enhancement of gas production in the reactor and at the same time the degradation rate indicating the efficiency of the treatment, what retention time the reactors ought to be designed to. Figure 6 shows the average methane production determined for the treatment period (specific methane production referred to time and volume unit) depending on the duration of the treatment.

Figure 6 clearly shows the differences between the sewage sludge and the biowaste containing VFG, too. At the biowaste, as a result of the higher proportion of the easily degradable organic material due to the VFG, with the reduction of the retention time from 30 to 10 days, the gas yield grew in the case of 1:2 and 1:3 seeding ratios. At the 1:1.5 seeding ratio, because of the initial unfavourable conditions (pH, hydrogen), this effect occurred later between day 20 and 40. Because of the sludge being less degradable, the methane production gradually increased until day 40.

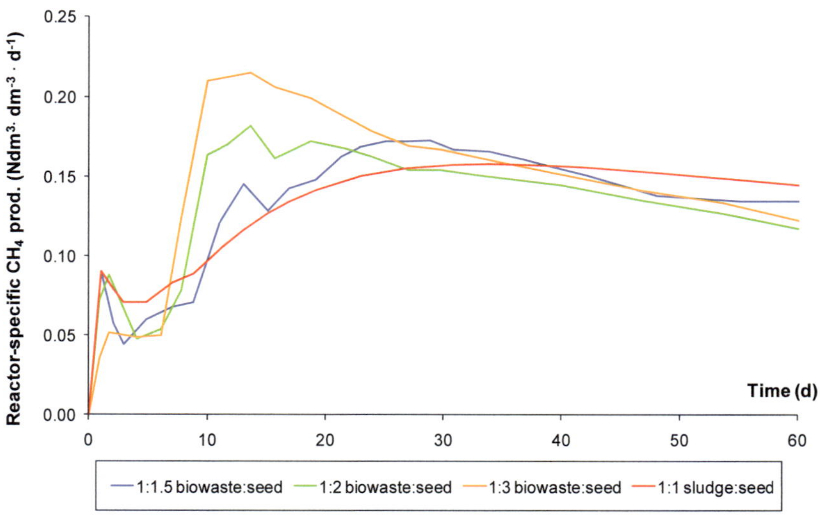

Fig. 6. Average methane production in the case of different biowaste : seed ratios

Figure 6 shows that in each case of waste to seed mixture, the average methane production reaches its maximum after day 10 and then after day 30 it starts to decline. This means that the retention time has to be minimum 30 days in the case of a combined dry batch treatment of VFG waste and sewage sludge. In the case of higher seeding ratios, following 30-40 days, the average methane production is almost the same in the case of each seeding, so thus, the effect of the seeding prevails less. Figure 6 confirms, that optimizing the anaerobic treatment, it is worth to check, together with the increase of the seed ratio, the option of reducing the retention time. It can be stated that the application of the seeding in 1:3 ratio has no negative impact on the gas production of the reactors even above a 40-day retention time assuring high grade stabilization.

4. Conclusion

Based on the test results we stated that the sewage sludge can be well degraded also through co-digestion by dry batch treatment together with VFG waste. We stated that in the case of 1:0.5 and 1:1 biowaste to seed ratios, the reactors became acidified. Even in the case of higher seeding ratios 9-day initial „lag" phase can occur. The hydrogen content of the biogas and the pH in the reactors indicate an initial accumulation of fatty acids in the reactor. We measured the highest, 54% organic material degradation in the case of 1:3 biowaste to seed ratio. Comparing our measurement results with literature data, it can be stated that the total methane production projected to one unit of organic material and the organic material degradation is nearly the same in total. Our laboratory-scale experiment, however, was influenced by the relatively long „lag" phase. Based on our tests it can be stated that it is a complex task to determine the optimal seeding ratio and retention time where a universal value cannot be given. In practice, the optimal values have to be determined one by one, taking into consideration the degradation target together with the specific gas yield projected to the reactor. From the aspects of costs reduction regarding the investment and operation, based on the values of the gas productions referred to the reactor volume, the 1:1.5 biowaste to seed ratio seemed to be the most efficient. This lets the conclusion be drawn that it is not worth to recycle the seeding material in the reactors in a higher ratio than this. According to the above, the compromising waste to seed ratio taking into consideration the different aspects is minimum 1:1.5 which takes into account the higher degradation of the organic material, as well as the quantity of the methane producible from one unit of reactor volume and the demand for low investment costs.

In the case of the same seed ratio, we experienced great difference in the efficacy of the treatment in case of biowaste containing VFG and the sewage sludge. During a co-digestion of sewage sludge and VFG wastes, because of the VFG waste having a quality varying in space and time, it is advisable to determine the suitable seed ratio through degradation tests in advance.

5. Acknowledgment

This research was funded by the Scientific Fund of the Eötvös József College.

6. References

Batstone, D. J., Keller, J., Angelidaki, R. I., Kalyuzhnyi, S. V., Pavlostathis, S. G., Rozzi, A., Sanders, W. T. M., Siegrist H., & Vavilin V. A. (2002). *Anaerobic Digestion Model No.*

1 (ADM1) Scientific and Technical Report No. 13 IWA Task Group for Mathematical Modelling of Anaerobic Wastewater IWA Publishing, London, England, pp. 88.

Benedek, P., Major, V., Réczey, G., & Takács I. (1990). *Biotechnológia a környezetvédelemben,* Műszaki Könyvkiadó, ISBN 963-10-8224-5, Budapest, Hungary

Biotechnion (1996): *Laboratory methods and procedures for anaerobic wastewater treatment,* Wageningen Agricultural University, Department of Environmental Biotechnology, The Netherlands

Breure A. M., & van Andel J. G. (1984). *Hydrolysi´s and acidogenic fermentation of a protein, gelatine, in an anaerobic continuous culture,* Applied Microbiology Biotechnology 20. pp. 40-45

Brummeler ten, E., Horbach H. C. J. M., & Koster I. W. (1991). *Dry anaerobic batch digestion of the organic fraction of municipal solid-waste,* Journal of Chemical Technology and Biotechnology, Volume: 50 Issue: 2, pp. 191-209.

Brummelet ten, E., Aarnink M. M. J., & Koster I. W. (1992). *Dry anaerobic-digestion of solid organic waste in a BIOCEL reaktor at pilot-plant scale,* Water Science and Technology, Volume: 25 Issue: 7, pp. 301-310.

Brummeler ten, E. (1993). *Dry Anaerobic Digestion of the Organic Fraction of Municipal Solid Waste,* Doctoral Thesis, Wageningen Agricultural University, Wageningen, The Netherlands.

Brummeler ten, E. (2000). *Full scale experience with the BIOCEL process.* Water Science and Technology: Vol. 41 (3) pp. 299-304.

Cout, D., Gennon, G., Ranzini, M., & Romano, P. (1994). Anaerobic co-digestion of municipal sludges and industrial organic wastes. In: *Proceedings of the 7th International Symposium on Anaerobic Digestion.* Johannesburg, South Africa.

Ghosh, S., & Klass, D. L. (1978). *Two-phase anaerobic digestion.* Process Biochemistry 13(4) pp. 15-24.

Gujer, W., & Zehnder A. J. B. (1983). *Conversion processes in anaerobic digestion.* Water Science and Technology 15. pp. 127-167

Hanaki, K., Matsuo T., & Nagase M. (1981). *Mechanism of inhibition caused by long-chain fatty acids in anaerobic digestion process,* Biotechnology Bioengineering, 23. pp. 1591-1610

Haug, R. T. (1980). *The Compost Engineering. Principles and Practice,* Ann Arbor Science Publishers, ISBN 0-250-40347-1, Michigan, The USA

Jeris, J. S., & McCarty, P. L. (1965). *The Biochemistry of Methane Fermentation Using 14C tracers,* Journal WPCF 37, No. 2 . pp. 178-192.

Kaspar, H. F., & Wuhrmann, K. (1978). *Kinetic parameters and relative turnover of some important catabolic reactions in degesting sludge.* Appl. Environ. Microbiol.

Kayhanian, M., & Tchobanoglous, G. (1992). Pilot investigation of an innovative two-stage anaerobic digestion and aerobic composting process for the recovery of energy from the organic fraction of MSW. In: *Proceedings of the 5th International Symposium on Anaerobic Digestion.* Venice, Italy.

Koster, I. W. (1989). *Toxicity in anaerobic digestion with emphasis on the effect of ammonia, sulfide and long-chain fatty acids on methanogenesis,* Doctoral dissertation, Wageningen Agricultural University, Wageningen, The Netherlands

Lettinga, G. & Hulshoff Pol, L. W. (1990). Basic aspects of anaerobic wastewater treatment technology. In: *Anaerobic reactor technology, International Course on Anaerobic Waste Water Treatment,* Wageningen Agricultural University

Lettinga, G., Zeeuw, W. J., & Ouborg, E. (1981). *Anaerobic treatment of wastes containing methanol and higher alcohols,* WaterRes., 15. pp. 171-182

Robinson, J. A., & Tiedje, J. M. (1982). *Kinetics of hydrogen consumption by rumen fluid, anaerobic digestor sludge and sediment,* Applied Envoronmental Microbiology, 44. pp. 1374-1384

Rózsáné, Sz. B., Simon, M., & Füleky, Gy. (2011). Effect of anaerobic pretreatment by dry batch technology on eerobic gegradability of sewage sludge, In: *Anaerobic Digestion: Processes, Products and Applications,* Nova Science Publishers, ISBN 978-1-61324-420-3. In press

Simon, M. (2000). *Települési szerves hulladékok kezelése és hasznosítása,* PhD értekezés. Debreceni Egyetem, Agrártudományi Centrum, Mezőgazdaságtudományi Kar, Környezetmérnöki Intézet, Víz- és Környezetgazdálkodási Tanszék, Debrecen

Tchobanoglous, G., Theisen, H., & Vigil, S. (1993). *Integrated Solid Waste Management.* Chapter 9, mcGraw-Hill, New York.

Wolin M. J., & Miller, T. L. (1982). *Interspecies hydrogen transfer: 15 years later,* ASM News, 48. pp. 561-565

Zehnder A. J. B., Ingvorsen, K. & Mari, T. (1982). Microbiology of methane bacteria, In: Hughes, D. E. at al. (eds.) *Anaerobic Digestion 1981,* Elsevier Biomedical Press, Amsterdam, The Netherlands, pp. 45-68

MSZ 21976/10:1982 Települési szilárd hulladékok vizsgálta, Kémiailag oxidálható szervesanyag-tartalom meghatározása

Part 2

Landfill and Other General Aspects of MSW Management

7

Separate Collection Systems for Urban Waste (UW)

Antonio Gallardo, Míriam Prades,
María D. Bovea and Francisco J. Colomer
*Universitat Jaume I, Castellón,
Spain*

1. Introduction

Separate collection of urban waste can be defined as a specific collection system that allows recoverable materials in waste to be separated. The human factor is very important in this new collection model, as the citizen now plays an active rather than a passive role as a processor of materials at source.

Various solutions for separate collection have emerged in order to fulfil all the objectives stipulated by legislation and local authorities in various countries for the recovery of municipal waste. Germany uses the Dual System, in which packaging waste is collected separately by a network belonging to companies selling consumer products. Separation of the organic fraction is mandatory in the Netherlands. In France and Spain, the governments are the responsible for designing mechanisms to implement the separate collection of packaging waste and achievement of the goals stipulated by European legislation. In the USA and Canada in the early 1990s, many cities with residential areas containing single-family homes began pilot schemes for separate collection, with waste separated at source into two, three and four fractions.

The large number of factors involved in establishing a separate collection system (economic, social, environmental, legal, etc.) means that there is no single solution or alternative. This has given rise to studies of citizens' behaviour with regard to the various collection systems: the level of participation, quality of the waste collected, financial incentives, etc. (White *et al.*, 1995; Wang *et al.*, 1997, Gallardo, 2000, Martin *et al.*, 2006, Shaw *et al.*, 2006, Dahlén *et al.* 2007). Other authors have analysed the various demographic, logistic and economic factors influencing citizens' participation, and assessed the quantities collected, generation and composition data for certain indicators based on these factors (Daskalopoulos *et al.* 1998, Emery *et al.*, 2003, González-Torre & Adenso-Díaz, 2005).

Separate collection of organic waste has been implemented in countries in northern Europe for several years, and is now relatively well established there. It is not yet widespread in Spain or others countries in southern Europe, but there are experiences in many autonomous regions at local or regional level which have had varying degrees of implementation and success. The Framework Directive on Waste (Directive 2008/98/EC) introduced the concept of biowaste and the need to recover this type of waste. The organic fraction of urban waste is considered biowaste, and it accounts for 36% (UE, 2011) of urban waste. As such, its separate collection is expected to be enhanced in all European Union countries, in order to use it and minimize its deposit into landfill.

2. Legislation

When planning the management of municipal solid waste in a particular area or region, the various actions and initiatives must be arranged hierarchically in accordance with the needs imposed by the environment. The hierarchy thus established will essentially depend on the policies of each region or state at a particular moment in time. Under Directive 2008/98/EC, European Union member states must observe the following hierarchy in waste management; it can be also used as a list of priorities for legislation and policies concerning waste prevention and management:

- Prevention
- Preparing for re-use
- Recycling
- Other recovery, e.g. energy recovery
- Disposal

However, the Directive also states that when this waste hierarchy is applied, the member states shall take steps to foster the options that provide the best overall environmental outcomes. This could mean that certain waste streams have to be removed from the hierarchy when a life cycle approach to the global impacts of waste generation and management calls for it.

As far as the evolution of the legislation on packaging and packaging waste is concerned, the recovery targets in the first Directive passed on this subject (Directive 94/62/EC) were:

- Recovery of 50-65%, by weight, of packaging waste before 2001.
- Recycling of 25-45%, by weight, of packaging waste, with a minimum of 15% for each material, before 2001.

The following Directive (Directive 2004/12/EC) is far stricter and raised the recovery targets to be achieved in 2008, which now stand at:

- Recovery: minimum 60% by weight (includes incineration with energy recovery)
- Recycling: between 55-80%, by weight, of the total amount of packaging waste. With the following minimum values:
- 60% by weight of glass
- 60% by weight of paper/cardboard
- 50% by weight of metal
- 22.5% by weight of recyclable plastic
- 15% by weight of wood

The progressive changes introduced into the legislation have meant that town councils have also had to evolve in terms of the separate collection methods and programmes they use in order to adapt to the new limits established by the legislation.

3. Urban waste separate collection systems

Separate collection is part of the comprehensive urban waste management system, which covers everything from collection to disposal (Tchobanoglous *et al.*, 1994), but which can be studied as a separate subsystem consisting of the pre-collection and collection stages (Figure 1). When studied separately, it can be considered an independent system, in which the input consists of a stream of urban waste and the output is several streams of selected materials, which go to the next management stage where it is subjected to different recovery methods. Separate collection is influenced by a number of environmental factors that influence the choice of the alternatives in the two elements of the system.

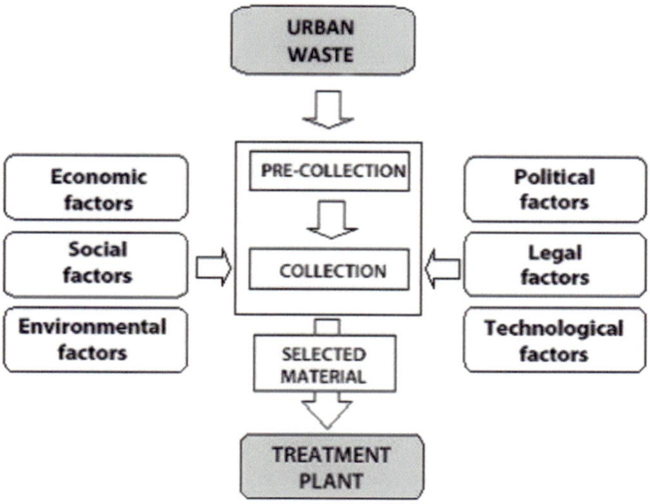

Fig. 1. Separate collection diagram.

Pre-collection includes handling, processing and storage of urban waste by citizens before it is deposited at the collection points, where there are different types of containers (Figure 2). After it has been deposited and stored at these points, the waste is collected and transported to the next facility in the urban waste management system. Most recovery methods, such as recycling or incineration, require separation into different fractions at source in order to achieve the minimum quality levels required in these processes. To that end, there is a wide range of fractioning at source, which can range from level zero, i.e. unfractioned or "all in one" collection, to a high level of specific separation of materials (Table 1).

Fig. 2. Urban waste collection containers in Spain.

Without fractioning	Total urban waste 
Two fractions: organic and mixed waste	Mixed Organic 
Three fractions: organic, lightweight and mixed waste	Mixed Organic Lightweight 
Specific separation: mixed and specific waste	Mixed Specific 

Table 1. Types of fractioning at source.

Once the waste has been separated, the question is what to do with it until it is collected. It is usually stored at home or at collection points located on the street (Figure 2). There are several levels of storage, depending on the distance travelled by the citizen to the deposit point:

- *The door to door system.* The bins or containers are located at each door, courtyard or other area accessible from the home or building. The distance that the citizen has to travel to deposit the waste is minimal.
- *The kerbside system.* The deposit points are no longer located at the door, but every 50-60 m on kerbsides. Citizens do not have to travel very far and acceptance is good. This system is applied in cities with high population densities.
- *The drop-off points system.* Collection points are located at greater distances in order to reduce management costs. The areas may have a range of between 100 and 300 m. The system relies on citizens' willingness to travel longer distances on foot. Figure 3 shows two drop-off points for dropping off paper-cardboard, glass and lightweight packaging (plastic, metal and liquid packaging board) separately.
- *Deposit at establishment level.* Some establishments (shops, public institutions, etc.) collaborate in the separate collection of some types of waste, particularly hazardous waste such as batteries, fluorescent lamps and drugs.
- *Deposit at facility level.* Deposit points are located in facilities away from the residential area. These facilities are called "Clean points", "Ecoparks" and "Recovery and Recycling Centres" (Household Waste Recycling Centres in the United Kingdom). They are able to selectively collect all types of waste, especially those that are not collected at other levels. An ecopark in Spain is shown in Figure 4.

Combining the different types of separation at source with storage levels can create a wide range of pre-collection systems. At one extreme is "all in one" pre-collection and the "door to door" level, which is the most convenient alternative from the citizen's standpoint. The managing agent is responsible for separation and recovery, which is more expensive and very inefficient in terms of recovery of materials. At the other extreme is collection with specific fractioning (paper, glass, packaging, batteries, etc.) at establishment and facility level. In this case, citizens have greater responsibility for separation, and materials with added value for the manager are obtained.

Fig. 3. Collection at drop-off points in Switzerland (left) and Spain (right).

Fig. 4. Ecopark or clean point in Spain.

3.1 Efficiency indicators

In order to determine the level of efficiency of a separate collection system, it is necessary to define some indicators. This efficiency has been defined in terms of the level of recovery of clean materials at source, deposited in the container, which is in turn expressed in terms of a series of indicators (Gallardo *et al.* 2010):

- *Fractioning Rate* (FR_i): the ratio between the amount by weight of the raw material separated and the total amount of urban waste. This rate is used to measure the various collection streams.

$$FR_i = \frac{Gross\,amount\,of\,waste\,collected\,in\,container\,for\,i}{Total\,amount\,of\,urban\,waste} \cdot 100(\%)$$

Where i is p, g, lp, o or uw depending on whether the FR is for paper, glass, packaging, organic waste or household waste, respectively. Raw material refers to material contaminated to a greater or lesser extent by other unwanted materials (also called improper materials).

- *Separation Rate (SR_i):* the ratio between the amount by weight of raw material separated and the total amount of material in the urban waste.

$$SR_i = \frac{Gross\,amount\,of\,waste\,collected\,in\,container\,for\,i}{Total\,amount\,of\,i\,waste\,generated} \cdot 100(\%)$$

For example, to find out the SR of the paper-cardboard in the paper container, we divide the amount of separated paper by the total paper in the waste.

- *Quality in Container Rate* (QCR_i): the ratio between the amount of net recyclable materials deposited in a container and the gross amount deposited in them.

$$QCR_i = \frac{Amount\,of\,waste\,collected\,correctly\,in\,container\,for\,i}{Gross\,amount\,of\,waste\,collected\,in\,container\,for\,i} \cdot 100(\%)$$

A good system would be one with a high SR with high quality materials, i.e. a low proportion of improper materials. These indicators are a necessary tool for evaluating the efficiency of the infrastructure of the separate collection system implemented. They provide direct information on the total quantities and products obtained in each type, in comparison with the potential amount of recyclable materials present in urban wastes. They are also useful for a diagnosis of the situation of the separate collection programme implemented and for finding out whether the poor quality of the materials collected separately is due to low participation, poor performance in collection or a lack of public information, for example.

3.1.1 Factors affecting the separation rate

The SR is directly related to the public's participation in separate collection programmes. This participation depends on several factors, which Wang *et al.* (1997), grouped into two categories: factors related to the citizen's attitude and factors associated with the characteristics of the collection programme. Few studies have been carried out in this area (Noehammer *et al.*, 1997), although some researchers have found interesting results that relate the SR to various factors:

- *The economic factor*. Noehammer *et al.* (1997), in a study on the impact of free separate collection bins on the degree of participation, concluded that providing bins in voluntary programmes has a positive effect. However, they were unable to confirm anything for compulsory programmes. They also showed that if the fee paid directly for the collection of municipal waste falls, the level of participation increases. Gilnreiner (1994) presents the results of a survey conducted in Vienna in which "reward systems" are clearly preferred by consumers over "punitive systems" such as ecotaxes and packaging taxes. Another way to increase the degree of participation is through the introduction of tax benefits (Bolaane, 2006) or by lowering the fees paid for collection (Harder & Woodard, 2007)
- *Size of housing*. Participation in high-rise housing, which is usually small in area, is lower than in single-family homes, which are generally larger. On the other hand, it is

easier to monitor who is participating in separation in the latter, while in a block of flats this is more difficult. Following this argument, in the Netherlands it has shown that occupants of tall buildings are more reluctant to participate (White *et al.*, 1995).

- *Frequency of collection.* If collection is infrequent, citizens' motivation declines, while if frequency is appropriate and in line with the rate containers are filled, participation increases (White *et al.*, 1995).
- *Number of fractions.* The level of participation falls as the number of fractions into which urban household waste is divided increases. Noehammer *et al.* (1997) studied the influence of the number of separations on the degree of participation in 104 separate collection programmes and found that for a separation into two fractions, participation is between 75-95%, while for more than four fractions the range was 49 -92%. Another conclusion they obtained was that in mandatory programmes, there was no clear correlation between the level of participation and the number of separations. However, when the programme was voluntary, the level of participation declined as the number of fractions increased.
- *Distance to the deposit point.* Participation falls as the distance to the deposit point increases. In Spain, the *SR* for glass in kerbside collection was 40% (distance to container 50 m), while at drop-off points, with a range of between 100-160 m, the average is 22% (Gallardo *et al.*, 1999).
- *Compulsory separate collection.* Compulsory separate collection programmes have a higher level of participation than voluntary programmes, providing that they are accompanied by certain incentives. Noehammer *et al.* (1997) found in their study that in compulsory programmes the level of participation had a range of 49-100%, while in voluntary collection it ranged between 11% and 92%, and concluded, as did the other authors that they cite, that compulsory programmes are more successful provided that they are accompanied by a high level of information, financial incentives, an adequate collection frequency, free containers, etc.
- *Socioeconomic level.* Gandy (1994) and Belton *et al.* (1994), in two studies on the relationship between socioeconomic level and the degree of participation in the drop-off points, found that there was participation clearly increased among people with a higher socioeconomic level. However, Lober (1996) in a study of reduction at source and recycling, found that in the various recycling programmes he studied, the socioeconomic factor was not significantly correlated with the degree of participation.
- *Education and promotion.* The level of information received by the public influences the degree of participation and quality of the separated materials. Gallardo (2000) showed that in Spain the *SR* of glass and paper-cardboard at the drop-off point level was higher in cities where citizens were better informed, for the same range.
- *Sociodemographic characteristics.* These characteristics have become a basic tool for many researchers when evaluating the profiles of a participatory individual compared to another person who does not participate in collection programmes. This classification also makes it possible to take education and awareness-raising action among groups with low participation (Rojas *et al.* 2008). Nationality, socioeconomic status, age and gender, among other factors, correlate strongly with separation behaviour (Rojas *et al.*, 2008). In Preston (England), Perry & Williams (2006) conducted research on participation in separate collection programmes, including the factor of nationality, because ethnic minorities now constitute a significant proportion of the town's

population. The study found that nationality that anticipated most in the recycling programme was the Indian-British minority (95.2%), followed by the British (78%) and the category of "other ethnic groups" (56.3%). The authors note that the reasons for the differences shown are unclear and require further investigation.

3.1.2 Factors affecting the quality in container rate

High levels of contamination can lead to considerable variations in the real quantities of material collected separately, its quality and its market price. To improve the *QCR* in pre-collection, the design of educational and social awareness campaigns and initiatives may be helpful. The input of contaminating materials may be due to different causes, of which the most common are:

- Deposit of waste in the wrong container.
- Deposit of the correct material but in the wrong format, such as depositing plastic toys in a container for plastic packaging.
- Deposit of dirty material, such as glass packaging with metal caps.

As with the *SR*, the *QCR* is also affected by numerous factors. The most important of these are:

- *The number of products separated together.* The fewer the different materials deposited in the same container, the higher the *QCR* achieved. The highest levels are reached in specific or single-material containers. Paper-cardboard containers in Spanish cities can achieve a QCR_p of 99.5% (Gallardo *et al.*, 2010). Kelleher (1996), in a study on the *QCR* in door to door collection with four fractions (organic, glass, paper-cardboard and packaging) found that the packaging fraction (plastic, metal and cardboard packaging) had 35% of improper materials while the paper fraction only had 2%.
- *Degree of complexity in separation.* Citizens are more sensitive to generic concepts than specific details, and as such the more specifications asked for in the separation, the lower the *QCR* obtained. This was confirmed in the Mancomunidad (Commonwealth of Municipalities) of Pamplona (Spain) (MCP 1997), which specified what type of plastic had to be put in the packaging container, and only 34% of the plastic was in fact the type specified.
- *The pre-collection characteristics.* When materials are presented in open bins, contamination levels are lower (5-8%) as the fact that the containers are open means that their contents can be inspected. When closed containers or bags are used, contamination levels rise to 27-36% (White *et al.*, 1995). Ayerbe (2000) presented a study conducted in a Spanish city which compared the quality of packaging collected in 1100 liters containers with an open lid and a closed lid (a system in which the container has an opening the size of a rubbish bag) and showed that in the former, the amount of improper materials (QCR_{lp}= 50%) was much higher than in the latter (QCR_{lp}= 74%)
- *Degree of public information.* Information campaigns have a major effect on the *QCR*. Various experiences have shown that if clear guidance is given to citizens on how to make the selection, they separate the different categories of waste perfectly. In the case of the Commonwealth of Pamplona (MCP, 1997), when there is separate collection in two fractions (recyclable materials and mixed waste), in 1995 the rural area had a *QCR* in recyclable material containers of 58%. A year later, after an information campaign

designed specifically for them, this had increased to 64%. According to Kimrey (1996), the information-education programme must be targeted at the entire family as if not, it is possible that not all members will collaborate in separate collection.

4. Separate collection systems in Europe

Since there are new European standards that must be complied with in terms of recycling targets, local authorities in countries belonging to the European Union have hastened to develop new collection models. As a result, a wide variety of separate collection systems can be found throughout Europe. This has given rise to their study and comparison in the different countries where they have been implemented. For example, in Sweden, Dahlén *et al.* (2007) conducted a study that compared 3 systems. The first consists of kerbside collection of recyclables and organic waste, including a specific case in which fees are paid for the collection of mixed waste. The second involves kerbside collection of recyclables, and the third the collection of recyclables at drop-off points. After the study, it was concluded that in municipalities with a kerbside collection system for recyclables, the mixed waste container has a higher *QCR* due to the proximity of the containers to the public. The introduction of the fee payment system for the collection of mixed waste led to this waste from many homes being burnt or dumped in the wrong place. In fact, in the municipality where this system was in place, the level of improper materials was 12% compared to 4% in one of the municipalities where this model was not implemented.

Mattsson *et al.* (2003) produced a study that compared separate collection systems in both single family and multifamily property close areas in Sweden and England. They used 6 Swedish examples and one recently introduced case in the United Kingdom as examples. The cases analysed showed that although the technical details were almost identical, they differed in terms of how they had been developed. The aspects taken into account for their different developments were: cooperation between the municipality and producers, the efficiency in collection using appropriate vehicles, the quality of the materials collected, Agenda 21 and environmental awareness.

In the United Kingdom, Woodard *et al.* (2001) analysed waste collection in a district before and after the introduction of a new plan. Under the old system, mixed waste was collected in black bins and recyclables (paper, cardboard and metal) in a box, every week in both cases. The new plan (CROWN: Composting and Recycling our Waste Now) added a green container for the collection of biowaste every two weeks. The frequency of mixed waste collection declined and became fortnightly, like biowaste. The container volumes and satisfaction of citizens was noted in a sample of households in a residential area. The result was a 55% reduction in waste sent to landfill sites, an increase from 5.5 to 17.7 liters in the average amount of deposited recyclables per household per week, and a rise in participation from 40% to 78%. Wilson and Williams (2007) subsequently analysed the implementation of a new collection system in a northern town in the United Kingdom. Two samples with different collection frequencies were used to see which system worked better: in one the mixed waste and recyclables were collected in alternate weeks, and in the other the mixed waste was collected weekly and the recyclables fortnightly. The proportion of containers brought to the street for collection compared to the total number of containers available was analysed in each sample, as was the level of material recovery in each sample. Both calculations produced better results in the first sample.

In Cappanori (Italy) four fractions are collected door to door: organic, multi-material, paper-cardboard and mixed waste. All the fractions are collected by the same truck on different days, so that each time one fraction is collected. In this way, you save considerably on transport (Connett, 2011).

In Portugal, Gomes *et al.* (2008) produced an economic comparison of three alternatives in terms of the collection of biowaste. The alternatives were: collection of biowaste without separation, separation of biowaste in the whole municipality and the separation of biowaste in the main urban centres and home composting in the rest. The costs of collection and transportation in the three stages were quantified and it was found that compared to the first, which was the one used in the study area, in the second the costs would not necessarily be higher, and costs could even be lower with the third.

In Spain, Ayerbe and Pérez (2005) analysed three lightweight packaging collection systems. The first consisted of collection from drop-off points. The second consisted of kerbside collection, next to the mixed waste container, using open top containers. The third was the same as the second but used closed lid containers (The closed lid has a hole with the size of a rubbish bag). A comparative analysis was performed of *QCR*, the yield of the packaging selection plant (the ratio between packaging material entering and leaving the plant). As for the *QCR*, the worst system was the open top, which obtained a proportion of improper materials of over 50%. In terms of performance on the ground, the best system was the first (73.4%) followed closely by the closed lid system (68.7%). Berbel *et al.* (2001) carried out a similar study in Cordoba. They compared two types of collection of lightweight packaging: with a container exclusively for this type of waste and a container for inert waste (packaging and other inert materials). The results showed that more mixed waste was collected from containers with the second system. Gallardo *et al.* (2010) conducted a study to determine the separate collection systems in place in Spanish towns with over 50,000 inhabitants, and their efficiency. They found that there are four different systems (Figure 5) and the most common is separation into four fractions: paper-cardboard, glass, lightweight packaging (plastic, metal and liquid packaging board) and mixed waste: mixed waste is collected at kerbside, while the paper-cardboard, glass and packaging are collected at drop-off points. They also found the *FR* for the four models. The main difference was that in the *FR* for packaging, this

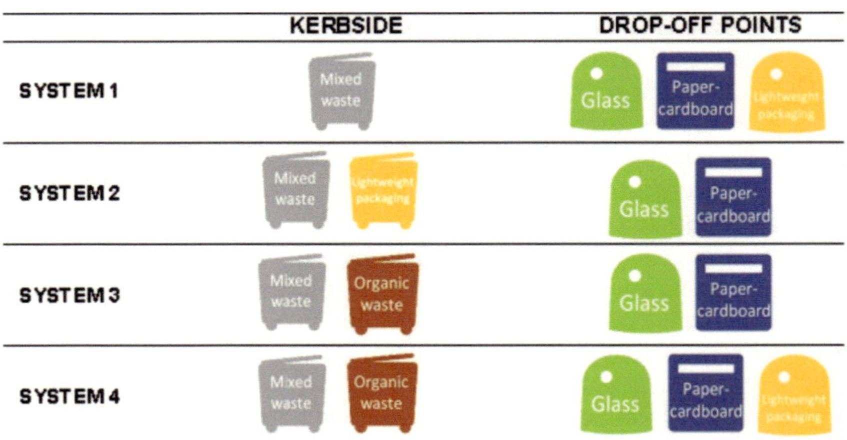

Fig. 5. Pre-collection models implemented in Spanish cities.

index is greater when lightweight packaging is deposited in kerbside bins (System 2), due to the fact that citizens have to travel a shorter distance to dump it. Furthermore, the materials left in System 2 contain a higher percentage of unwanted materials than when they are collected at drop-off points (System 1 and System 4), i.e. the QCR_{lp} is lower. This is due to the proximity to the mixed waste bin and to the fact that when it is filled to overflowing, citizens leave their waste in the lightweight packaging bin. In addition, in the collection of paper/cardboard, glass and lightweight packaging at drop-off points (System 1 and System 4), the *SR* of these fractions varied depending on the distance that people had to travel to deposit their waste; in specific terms, the greater the distance the lower the *SR*.

5. Separate collection systems for biowaste

An average of 520 kg of urban waste was generated *per capita* in the European Union in 2005 (Blumenthal, 2011) and this figure is expected to rise to 562 kg by 2020 (EU, 2011). If we consider this prediction to be correct, 290 million tons will be generated in 2020, of which 36% will be organic waste, and as such the amount would rise to 104.4 million tons. This is a very significant amount, and as such one would expect that the technology and the number of facilities for the use of this material would be greatly enhanced.

The organic fraction of urban waste consists of biodegradable materials (food scraps, spoiled food, gardening waste, etc.). European Union legislation, in the Framework Directive on Waste (Directive 2008/98/EC) defines "biowaste" as "biodegradable garden and park waste, food and kitchen waste from households, restaurants, caterers and retail premises and comparable waste from food processing plants", meaning that the organic fraction can be classified as biowaste, like wood, sewage sludge, and agricultural and forestry waste. This is very important from a legal standpoint, since this Directive requires EU member countries to implement separate collection and recovery, thereby reducing greenhouse gas emissions, the recovery of biowaste as biogas and compost, and a reduction in the amount of waste dumped at landfill sites.

The main objective of the separate collection of the organic fraction of urban waste is to convert it into high quality compost. This material can be used as fertilizer for agriculture, gardening or landscape restoration work. The critical factor is the percentage of improper materials accompanying the organic fraction of urban waste. It is therefore essential that this operation is carried out under the best possible conditions.

The greatest difficulty in establishing a separate collection system is designing the pre-collection, to make it as convenient as possible for citizens and not unduly expensive for the Council. Two aspects depend on public participation: the amount of waste collected and its quality. The former justifies the system's existence and the latter prevents composting centres from receiving waste that is more similar to the mixed waste fraction than the organic fraction, as occurs in some cases. Another important external factor that can affect the system is the existence of a potential market which guarantees the destination of the compost in the territory of the composting plant. If this market does not exist, alternatives such as biomethanation or incineration after drying can be considered. In order to minimize the energy costs of drying, biological drying systems (biodrying), solar drying or a combination of both can be used (Adani *et al.*, 2002, Velis *et al.*, 2009, Zhang *et al.*, 2009).

The systems that can be applied to the collection of the organic fraction of urban waste are the same as those mentioned above. As described above, special attention must be paid to the pre-collection, which consists of separation in an organic waste bin, preferably in compostable bags, such as those made of corn starch. The bags are then deposited in special containers for transport to composting plants. In order to minimize the amount of improper materials appearing in the containers, initial campaigns and information maintenance initiatives are carried out, and biodegradable bags are often given away. It is also quite useful to conduct a pilot test in one part of the city, in order to gain experience and roll out the service to the rest of the city.

In order to increase the amount collected and reach more generation points, the collection of material can also be arranged according to its origin (Marrero, 2010):

- Household (fruit and vegetable peelings, food scraps, fish scraps, meat bones and scraps, spoiled food, grass cuttings, small prunings, etc.).
- Restaurants, bars, schools and public buildings.
- Waste from markets, shops and services.
- Waste from parks, gardens and cemeteries.

After a separate collection system for the organic fraction of urban waste that meets the needs of the population concerned has been established, its main advantages are:

- It generates higher quality compost than the mixed waste fraction, which needs prior treatment.
- Reduced costs of subsequent treatment of the compost (compost refining).
- It minimizes the problem of landfill leachate.
- It complies with regulations currently in force.
- Biocompartmentalized containers are available, making savings during collection possible.

6. Case study: Efficiency of separate collection of the organic fraction in Spain

In order to meet European targets on waste (Directive 2004/12/EC) and comply with Spanish law, which requires councils in towns with over 5,000 people to implement separate collection systems, Spanish councils have had to design new collection models to adapt to these laws. For this reason, there is a wide variety of collection systems in Spain.

We present the results of a research paper which analyses separate collection systems for organic waste in Spanish towns with between 5,000 and 50,000 inhabitants. The systems and their efficiency are studied using the indicators *Fractioning Rate and Quality in Container Rate* and *Separation Rate.*

In the year taken into account in this study (2008), the population of Spain rose from 46,765,807 inhabitants in 2008 (INE, 2008a). In addition, according to data released by the Spanish National Institute of Statistics, 24,240,470 tonnes of municipal solid waste were collected in the year 2008, and the *per capita* collection rate was 611.82 kg (INE, 2008b).

The number of municipalities with between 5,000 and 50,000 inhabitants in 2008 was 1,145 (INE, 2008a). Studying this entire population would be a difficult task and would entail considerable time and effort. To that end, a representative sample of that population (279 towns) was defined according to a number of statistical variables. Each one was sent a survey by mail, requesting the following information:

- General information about the municipality: number of inhabitants, area and collection system in place.
- For each of the waste fractions collected separately: tonnes collected annually; composition; the year separate collection was implemented; the number of containers and frequency of collection.

After the entire information gathering process, data was available for 115 towns (41% of the towns in the sample), in 14 of the 17 Spanish regions.

Of all the towns for which information was available, 29.5% collect the organic fraction of urban waste, the majority are in the region of Catalonia, as the legislation there requires this type of collection. Such a low percentage is due to the fact that collection of the organic fraction of urban waste is still voluntary, and as such the majority of the towns have not yet implemented it. According to the study, there are 6 different collection systems, with the following characteristics:

- SYSTEM A: separation into 4 fractions (mixed waste, organic waste, paper-cardboard and glass). Mixed waste and biowaste is collected at kerbside, while paper-cardboard and glass are collected at drop-off points.
- SYSTEM B: separation into 5 fractions (mixed waste, organic waste, paper-cardboard, glass and lightweight packaging). Mixed waste and biowaste is collected at kerbside, while paper-cardboard, glass and lightweight packaging are collected at drop-off points.
- SYSTEM C: separation into 5 fractions (mixed waste, organic waste, paper-cardboard, glass and lightweight packaging). Mixed waste and biowaste is collected at kerbside, while paper-cardboard, glass and lightweight packaging are collected at drop-off points. The collection of biowaste is partially implemented and collected door to door. This is a variation on System 4.
- SYSTEM D: separation in 4 fractions (mixed waste, organic material, glass and multi-product[1]). Mixed waste and biowaste are collected at kerbside, while multi-product and glass are collected at drop-off points.
- SYSTEM E: separation in 4 fractions (mixed waste, organic material, glass and multi-product). Mixed waste, biowaste and multi-product are collected door to door, while glass is collected at drop-off points. This is a variation on System D.
- SYSTEM F: separation into 5 fractions (mixed waste, organic waste, paper-cardboard, glass and lightweight packaging). All fractions are collected at the kerbside.

The diagram of the 6 collection systems can be seen in Figure 6. Table 2 shows the towns that have implemented each of the systems above.

Table 2 shows how system B is used in most of the municipalities studied. There is a new fraction, multiproduct, in systems E and F, in order to optimize collection. This fraction is not very widespread, and is not found in large Spanish towns (Gallardo et al., 2010). Figures 7-12 shows the different FR_o obtained by each system and Table 3 shows the QCR_o and SR_o for organic waste.

[1] Multi-product: light packaging and paper-cardboard

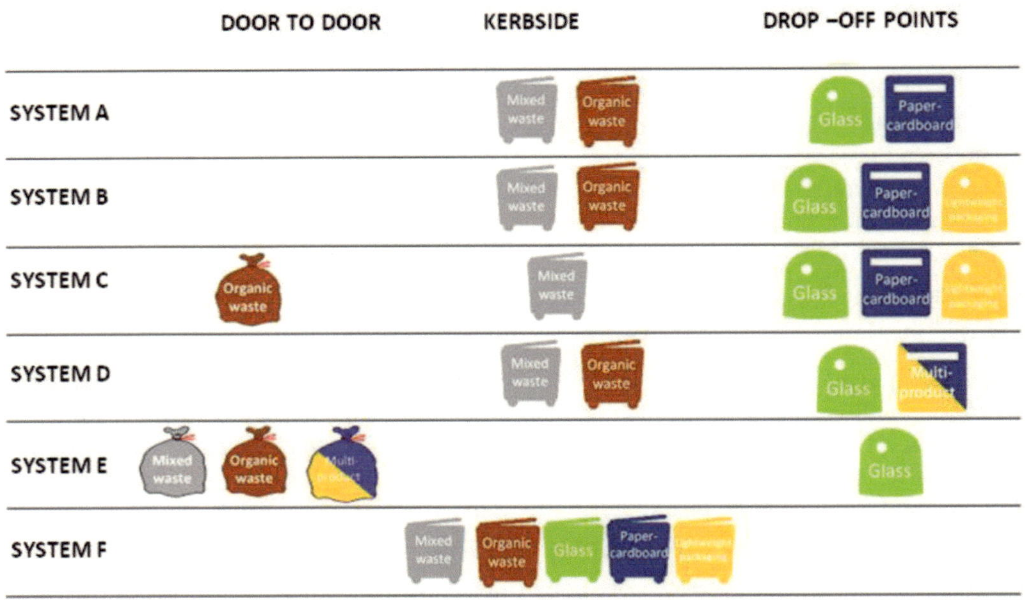

Fig. 6. Diagram of separate collection systems.

SYSTEM	No. cities
A	7
B	16
C	2
D	2
E	2
F	1

Table 2. Towns with between 5,000 and 50,000 inhabitants with each system.

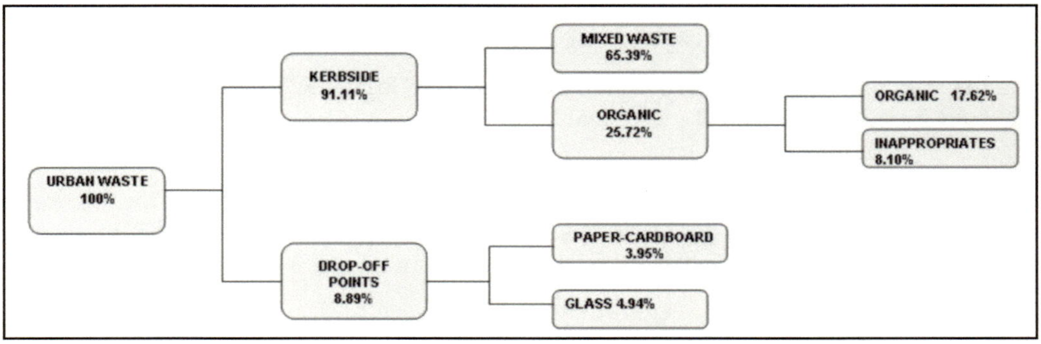

Fig. 7. System A Fractioning Rates.

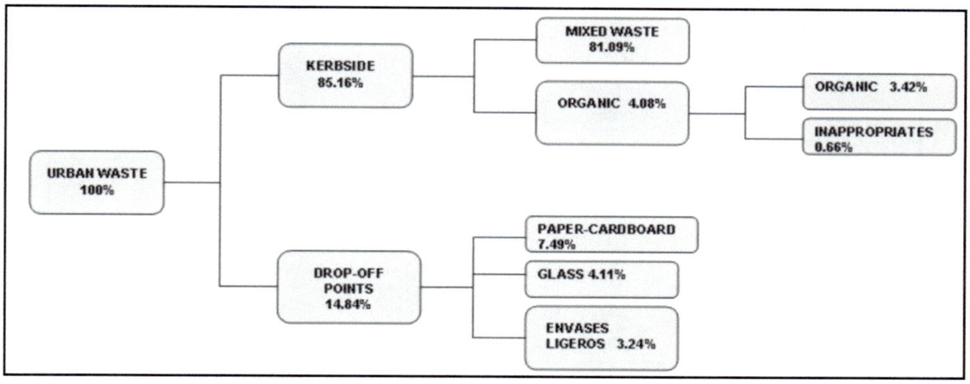

Fig. 8. System B Fractioning Rate

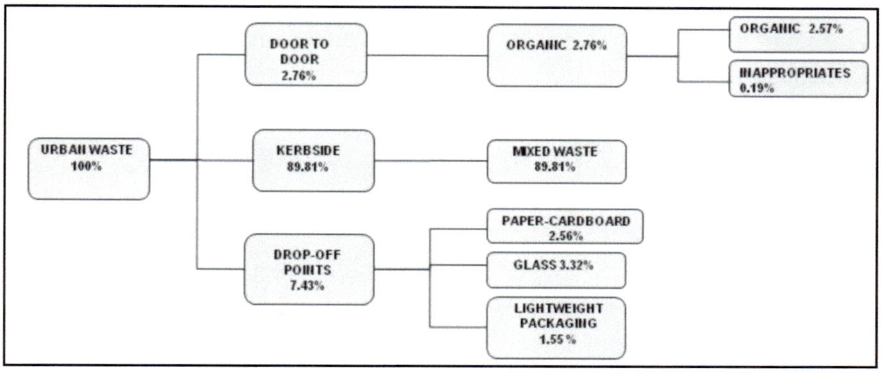

Fig. 9. System C Fractioning Rate

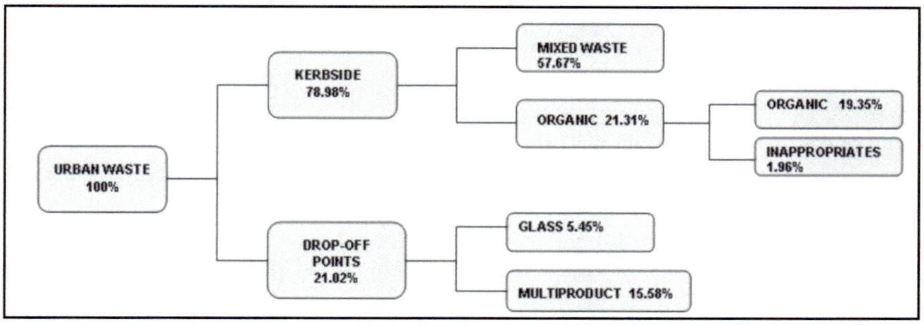

Fig. 10. System D Fractioning Rate

Using the FR_o and QCR_o calculated, it can be seen which system works best from the point of view of collection of the organic fraction of urban waste. The best FR_o results are obtained in system E, which also has the best QCR_o. The collection is door to door, which is very convenient for citizens, who do not have to travel any distance to deposit their waste. This system is suitable for towns in which the containers can be located inside buildings or homes. The worst FR_o and QCR_o results are for systems C and A respectively. The low FR_o is because the public participation is very low, as people prefer to deposit their waste in kerbside containers. Despite the low FR_o in system C, its QCR_o is high, which means that the few people

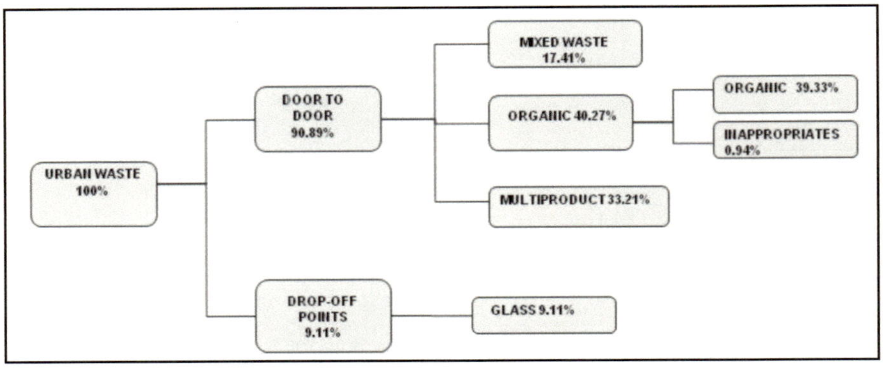

Fig. 11. System E Fractioning Rate

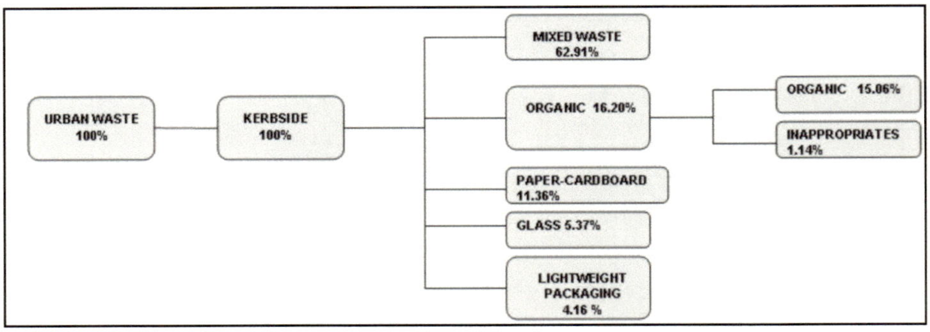

Fig. 12. System F Fractioning Rate

System	A	B	C	D	E	F
QCR_o (%)	68.51	83.82	93.12	90.80	97.67	92.96
SR_o (%)	71.51	24.50	12.92	33.44	76.22	37.85

Table 3. QCR_o and SR_o obtained in each system.

who do participate in this collection do it properly. The reason behind the low QCR_o in system A is the proximity to the mixed waste container, as if the mixed waste container overflows, or even in cases of confusion, mixed waste can be deposited in the organic waste container. The mixed waste container in system A contains approximately 40% of organic waste, meaning that information campaigns are required so that citizens are more aware of this type of collection.

Regarding the SR_o, it can be seen how system E has the highest value, which leads us to conclude that this is the best system. The proximity of the container to the citizen and a higher level of fractioning are undoubtedly factors in obtaining good results in the separate collection of organic waste.

7. References

Adani, F., Baido, D., Calcaterra, E., & Genevini, P. (2002) The influence of biomass temperature on biostabilization-biodrying of municipal solid waste. *Bioresource Technology*, Vol. 83, No. 3, pp. 173-179.

Ayerbe, S. (2000). La recogida selectiva: punto clave para el reciclado, *Proceedings of Seminario sobre cultura medioambiental en la gestión de los residuos urbanos*, Lorca (Spain).

Ayerbe, S., & Pérez, S. (2005). Alternativas para la recogida de envases ligeros, *Ingeniería Química,* No. 423, pp. 203-207.

Belton, V., Crowe, D.V., Matthews, R., & Scott, S. (1994). A survey of public attitudes to recycling in Glasgow (U.K.), *Waste Management and Research,* Vol. 12, No. 4, pp. 351-367.

Berbel, J., Peñuelas, J., Ortiz, J., & Gómez, M. (2001). Análisis comparado de modelos de recogida selectiva de envases/orgánico, *Residuos* No. 59, pp. 52-57.

Blumenthal, K. (2011). Generation and treatment of municipal waste. Eurostat 31/2011 – Statistics in focus.

Bolaane, B. (2006). Constraints to promoting people centred approaches in recycling, *Habitat International, Vol.* 30, No. 4, pp. 731–740.

Connett, P. (2011). Zero Waste: A Key Move towards a Sustainable Society. August 2011. Available from: http://www.americanhealthstudies.org/zerowaste.pdf

Daskalopoulos, E., Badr, O., & Probert, SD. (1998). Municipal solid waste: a prediction methodology for the generation rate and composition in the European Union Countries and the United States of America, *Resources, Conservation and Recycling,* Vol. 24, No. 2, pp. 155-166.

Dahlén, L., Vukicevic, S., Meijer J., & Lagerkvist, A. (2007). Comparison of different collection systems for sorted household waste in Sweden, *Waste Management* , Vol. 27, No. 10, pp. 1298-1305.

Emery, AD., Griffiths, A.J., & Williams, K.P. (2003). In a depth study of the effects of socio-economic conditions on household waste recycling practices, *Waste Management and Research,* Vol. 21, No. 3, pp. 180-190

Gallardo, A., Tejero, I., & Ferrer, J. (1999). Alternativas en la recogida selectiva ante el nuevo marco normativo, *Proceedings of VI Congreso de Ingeniería Ambiental*, Bilbao, February, 1999.

Gallardo, A. (2000). *Metodología para el diseño de redes de recogida de selectiva de RSU utilizando sistemas de información geográfica. Creación de una base de datos aplicable a España,* Universidad Politécnica de Valencia, Valencia.

Gallardo, A., Bovea, M., Colomer, F., Prades, M., & Carlos, M. (2010). Comparison of different collection systems for sorted household waste in Spain, *Waste Management,* Vol. 31, No. 7, pp. 379-406.

Gandy, M. (1994). A comparative overview of recycling in London and Hamburg, *Waste Management and Research,* Vol. 12, No. 6, pp. 481-494

Gilnreiner, G. (1994). Waste minimization and recycling strategies and their changes of success, *Waste Management and Research,* Vol. 12, No. 3,pp. 271-283.

Gomes, A., Matos, M., & Carvalho, I. (2008). Separate collection of the biodegradable fraction of MSW: An economic assessment, *Waste Management* , Vol. 28, No. 10, pp. 1711-1719.

González-Torre, P.L., & Adenso-Díaz, B. (2005). Influence of distance on the motivation and frequency of household recycling, *Waste Management,* Vol. 25, No. 1, pp. 5–23.

Harder, M.K., & Woodard, R. (2007). Systematic studies of shop and leisure voucher incentives for household recycling, *Resouces, Conservation and Recycling,* Vol. 51, No. 4, pp. 732–753.

INE, 2008a. Demografía y población. Cifras de población y censos demográficos, In: *Instituto Nacional de Estadística,* Available from: www.ine.es

INE, 2008b. Sociedad. Análisis sociales, In: *Instituto Nacional de Estadística,* Available from: www.ine.es

Kelleher, M. (1996). Four stream residential collection in Holland, *Biocycle,* October, pp. 46-50.

Kimrey, E. (1996). Rethinking the refuse/recycling ratio, *Biocycle,* July, pp. 44-47.

Lober, J. (1996). Municipal solid waste policy and public participation in household source reduction, *Waste Management and Research,* Vol. 14, No. 2, pp. 125-143

Martin, M., Williams, I.D., & Clark, M. (2006). Social, cultural and structural influences on household waste recycling: a case study. *Resources, Conservation and* Recycling, Vol. 48, No. 4, pp. 357–395.

Marrero, X. (2010). El 5º contenedor llaga a Vitoria-Gasteiz: el reto de los residues orgánicos. *Proceedings of 10 Congreso Nacional de Medio Ambiente (CONAMA).* Madrid, 2010.

Mattsson, C., Berg, P., & Clarkson, P. (2003). The development of systems for property close collection of recyclables: experiences from Sweden and England, *Resources, Conservation and Recycling,* Vol. 38, No. 1, pp. 39-57.

MCP (1997). *Los envases en la gestión integral de los residuos.* Mancomunidad de la Comarca de Pamplona (MCP). Pamplona

Noehammer, H.C., & Byer, P.H. (1997). Effect of design variables on participation in residential curbside recycling programs, *Waste Management and Research,* Vol. 15, No. 4, pp. 407-427

Perry, G.D., & Williams I.D. (2006). The participation of ethnic minorities in kerbside recycling: A case study, *Resources Conservation and Recycling,* Vol. 49, No. 3, pp. 308-323.

Rojas-Castillo, L.D., Gallardo, A., Aznar, P., Ull-Solis A., & Piñeros, A. (2008). La participación ciudadana en los sistemas de recogida selectiva de residuos urbanos, un factor clave en la gestión, *Proceedings of I Simposio Iberoamericano de Ingeniería de residuos,* Castellón (Spain), July, 2008.

Shaw, P.J., Lyas, J.K., & Hudson, M.D. (2006). Quantitative analysis of recyclable materials composition: tools to support decision making in kerbside recycling, *Resources, Conservation and Recycling,* Vol. 48, No. 3, pp. 263–279.

Tchobanoglous, G., Theisen, H., & Vigil, S.A . (1994). *Gestión Integral de Residuos Sólidos,* McGraw-Hill, Madrid.

UE (2011). Analysis of the evolution of waste reduction and the scope of waste prevention. European Commission. DG Environment. Framework contract ENV.G.4/FRA/2008/0112

Velis, C.A., Longhurst, P.J., Drew, G.H., Smith, R., & Pollard, S.J.T. (2009) Biodrying for mechanical-biological treatment of wastes: A review of process science and engineering. *Bioresource Technology,* Vol. 100, No. 11, pp. 2747-2761.

Wang F.S., Richardson, A.J., & Roddick, F.A. (1997). Relationships between set-out rate, participation rate and set-out quantity in recycling programs, *Resources, Conservation and Recycling,* Vol. 20, No. 1, pp. 1-17.

White, P.R. , Franke, M., & Hindle, P. (1995). *Integrated Solid Waste Management. A lifecycle Inventory.* Chapman & Hall, ISBN 0-8342-1311-7, New York

Wilson, C., & Williams, I. (2007). Kerbside collection: a case study from the north-west of England, *Resources, Conservation and Recycling,* Vol. 52, No. 2, pp. 381-394.

Woodard, R., Harder, M., Bench, M., & Philip, M. (2001). Evaluating the performance of a fortnightly collection of household waste separated into compostables, recyclates and refuse in south England, *Resources, Conservation and Recycling,* Vol. 31, No. 3, pp. 265-284.

Zhang, D-Q., He, P-J., & Shao, L-M. (2009) Sorting efficiency and combustion properties of municipal solid waste during bio-drying. *Waste Management,* Vol. 29, No. 11, pp. 2816-2823.

Utilization of Organic Wastes for the Management of Phyto-Parasitic Nematodes in Developing Economies

P.S. Chindo[1], L.Y. Bello[3] and N. Kumar[2]
[1]*Department of Crop Protection, Institute for Agricultural Research, Ahmadu Bello University , Zaria*
[2]*Department of Crop Production, Faculty of Agriculture, Ibrahim Badamasi Babangida University, Lapai*
[3]*Department of Crop Protection, Federal University of Technology, Minna*
Nigeria

1. Introduction

The agricultural system in Nigeria and most developing countries has been dominated by the use of inorganic fertilizers as nutrient sources and synthetic pesticides for the management of pests and diseases. However the prices of these agro-chemicals have been skyrocketing beyond the reach of the rural poor farmer. Associated with this is their availability which is very highly unpredictable, thereby exposing the farmers to undue hardships in the crop production chain. Due to the high prices and unpredictable nature of the availability of these inputs, the rural poor farmers have resorted to utilizing organic materials /wastes principally as nutrient sources. These wastes however, have been shown to control a number of pests and diseases.

The term' waste' can be loosely defined as any material that is no longer of use, useless, of no further use to the owner and is, hence discarded or unwanted after use or a manufacturing process. These materials include agricultural wastes in the form of farm yard manure and dry-crop residues, sewage sludge, municipal refuse, industrial by-products, such as oilcakes, sawdust and cellulosic waste. Others are animal wastes such as feathers, bone meal, horn meal, and livestock wastes. Most discarded wastes, however, can be reused or recycled. This is the basis of the rag picking trade, the rifting through refuse dumps for recovery and resale of some materials. Today, heaps of refuse dump sites are disappearing in Nigeria because farmers evacuate them for use on their farms as organic fertilizers. Fortunately, these have been found to control phyto-parasitic nematodes among other diseases (Abubakar and Adamu, 2004; Abubakar and Majeed, 2000; Akhtar and Alam, 1993; Chindo and Khan, 1990; Hassan *et al.*, 2010). This is becoming an unconscious but well organized economically important waste management practice in Nigeria and many West African countries with attendant environmental benefits.

In recent years, there has been tremendous increase in public awareness on environmental pollution and climate change associated with pesticide toxicity and residues. This resulted in the shift in pest control strategies from chemical to the environmental era in the late 1980s. Since then several workers have reported that waste materials either of animal, plant or industrial origin have nematicidal and plant growth promoting properties (Akhtar and Alam, 1993; Chindo & Khan, 1990; Kimpinski *et al.*, 2003. This has been exploited as an alternative means of nematode control (Abubakar and Adamu, 2004; Abubakar and Majeed, 2000; Hassan *et al.*, 2010; Nico *et al.*, 2004; Nwanguma and Awoderu, 2002;). The beneficial effects of organic incorporation have been generally considered to be due to increase in soil nutrients, improvement in soil physical and chemical properties (Huang and Huang, 1993; Hungalle, *et al.*, 1986; Kang *et al*, 1981), direct or indirect stimulation of predators and parasites of phyto-parasitic nematodes (Kumar, 2007; Kumar et al, 2005; Kumar and Singh, 2011), and release of chemicals that act as nematicides (Akhtar and Alam, 1993; Sukul, 1992). Very often, when there was a decrease in the soil-pathogen population, there was a consequent increase in crop yield. (Akhtar, 1993; Akhtar and Alam 1993; Chindo and Khan, 1990).

Given the high cost and unpredictable supply of inorganic fertilizers and synthetic nematicdes, the best way to overcome such condition in the developing economies is to utilize waste resources for sustainable crop production and plant disease management. Given the importance of organic wastes highlighted above, this chapter intends to:

i. put together the research works published on the utilization of organic wastes for the management of plant disease with special reference to phyto-parasitic nematodes,
ii. examine the prospects of their usage in modern day agriculture,
iii. look at the challenges posed in the utilization of these wastes particularly in large scale agriculture, and
iv. attempt to proffer suggestions towards addressing these challenges.

2. Deployment of organic wastes for the management of phyto-parasitic nematodes

The food and agricultural organization (FAO) of the United Nations defines sustainable agriculture as a practice that involves the successful management of resources for agriculture to satisfy human needs while, maintaining or enhancing the quality of the environment and conserving natural resources (FAO, 1989). The system does not unduly deplete the resource as it makes best use of energy and materials, ensure good and reliable yields, and benefit the health and wealth of the local population at competitive production costs (Wood, 1996). Organic wastes perfectly fit into this definition. Being products of crop farms, domestic use, animal or industrial wastes, they are often recycled from the soil to farm produce thereby ensuring conservation of resources and environmental cleanliness. In addition, indirect benefits of pest and disease management are achieved. Numerous examples of these benefits on the management of phyto-parasitic nematodes have been reported by several workers.

2.1 Wastes from plants and plant origin

Compost made of agricultural and industrial wastes have been widely used as amendment in soil for the management of soil-borne diseases (Hoitink and Boehm, 1999;

Shiau et. al., 1999). In particular, several authors have reported suppression of diseases caused by root-knot nematodes with composted agricultural wastes (McSorely and Gallaher, 1995; Oka and Yerumiyahu, 2002). McSorely and Gallaher (1996) reported reductions in populations of the nematodes Paratrichodorus minor, M.incognita, Criconemella spp and Pratylenchus spp following applications of yard waste compost on maize (Zea mays) in Florida, USA. Forage yield of maize was increased by 10 to 212% when compared with the control.

In Spain, Andres, *et al.* (2004) using different composted materials at different rates in potting mixtures for the management of *Meloidogyne* species, found that root galling and final nematode populations of *M. incognita* race1 and *M. javanica* in tomato and olive plants were reduced. Increasing the rate of the test materials exponentially reduced galling and final population density of *M.incognita* by 40.8 and 81.9%, respectively (Table 1). Similar results were obtained for *M. javanica*. In south western Nigeria, Olabiyi *et al.* (2007) found that both decomposed and un-decomposed manure applied as organic amendment caused significant reduction in the soil population of *Meloidogyne* spp. *Helicotylenchus* sp. and *Xiphinema* sp. on cowpea. The organic manure resulted in a significant reduction of root galls on the cowpea (Table 2).

Composted amendment		Tomato				Olive
		M. incognita race 1		*M. javanica*		*M. incognita* race 1
Material	Rate (%)	RGS[b]	Final population[c]	RGS	Final population	Final population
Dry cork		Experiment I				
	0	3.3[d] a	114,612 a	4.0 a	73,777 a	8379 a
	25	2.4 b	39,752 b	2.9 b	25,470 b	1565 b
	50	1.8 c	17,396 bc	2.1 c	16,138 c	573 c
	75	1.5 c	8435 c	1.1 d	3483 d	248 c
	100	1.5 c	1737 c	1.0 d	215 d	132 c
Dry-grape marc		Experiment II				
	0	4.1 a	59,480 a	4.3 a	61,490 a	
	25	3.1 b	42,926 b	3.2 b	43,946 b	
	50	3.2 b	39,144 b	3.3 b	40,164 b	
	75	3.3 b	41,144 b	3.3 b	42,164 b	
	100	3.2 b	44,296 b	3.3 b	45,316 b	
Dry-olive marc + dry-rice husk (1:1)						
	0	3.7 a	69,499 a	3.8 a	65,609 a	
	25	3.4 a	64,914 a	3.5 a	60,750 a	
	50	3.4 a	72,931 a	3.5 a	68,249 a	
	75	3.6 a	71,208 a	3.4 a	68,654 a	
	100	3.4 a	64,078 a	3.4 a	63,561 a	

[a]Plants were inoculated with 10,000 eggs + second-stage juveniles (J2s) (tomato) or 5000 eggs + J2s (olive) at the time of transplanting into the amended potting mixture. Plants were incubated in a growth chamber under conditions favorable for nematode development for 2 months.

[b]Severity of root galling (*RGS*) was rated on a 0–5 scale according to the percentage of galled tissue, in which 0 = 0–10% of galled roots; 1 = 11–20%; 2 = 21–50%; 3 = 51–80%; 4 = 81–90%; and 5 = 91–100%.

[c]Final nematode population determined by extracting nematodes from 100-cm^3 samples of infested soil mixtures and from 5-g root samples of each plant 2 months after inoculation.

[d]Data are the average of two trials each with seven replicated plants per treatment combination. Means followed by the same letter do not differ significantly ($P > 0.05$) according to Fisher's protected LSD test.

Source; Andres, *et al.* (2004).

Table 1. Effects of composited amendments of potting mixtures on the root galling and finl population of *Meloidogyne incognita* race I and *M. javanica* on tomato and olive planting stock.

Treatment	*Meloidogyne* spp. Initial population	Final population	% reduction	*Helicotylenchus* sp. Initial population	Final population	% reduction	*Xiphinema* sp. Initial population	Final population	% reduction	Gall index
Decomposed wild sunflower leaf	546	341b	62.45	113	38a	33.63	69	41a	59.42	1.2a
Decomposed maize stover	569	308a	54.13	108	47a	43.52	72	38a	52.78	1.0a
Decomposed cassava peel	581	311a	53.53	119	31a	26.05	75	35a	46.67	1.0a
Undecomposed wild sunflower leaf	552	361b	65.40	115	48a	41.74	64	31a	48.44	1.3a
Undecomposed maize stover	568	307a	54.05	109	29a	26.61	67	33a	49.25	1.0a
Undecomposed cassava peel.	580	402b	69.31	107	35a	32.71	62	40a	64.52	1.5a
Control	577	2641c	457.71	114	188b	164.91	70	227b	324.29	4.7a
	NS		NS	NS						

NS = Not Significant
Means followed by the different letter(s) along the same column are not statistically different at 5% probability level

Source; Olabiyi, *et al.*, 2007

Table 2. Soil Nematode Population in 200 ml soil sample at planting (initial population) and harvest (final population), percentage reduction of nematodes and root gall index.

2.2 Use of plant parts

Numerous plant parts used as organic amendments have been shown to control phyto-parasitic nematodes. Neem (*Azadirachta indica*) is the best known example that act by releasing pre-formed nematicidal constituents into soil. Neem products, including leaf, seed kernel, seed powders, seed extracts, oil, saw dust and particularly oil cake, have been reported as effective for the control of several nematode species (Egunjobi and Afolami, 1976, Akhtar, 1998). Neem constituents, such as nimbin, salanin, thionemone, azadirachtin and various flavonoids, have nematidal effects; triterpene compounds in neem oil cake inhibit the nitrification process and increase available nitrogen for the same amount of fertilizer (Akhtar and Alam, 1993).

Akhtar (1998) reported the effect of two neem-based granular products, Achook and Sunneem G, urea and compost manure incorporated in the soil. These treatments were found to decrease the number of phyto-parasitic nematodes with increasing doses of plant products. Combination of both neem with urea were reported to be the most effective in suppressing this pathogen.

2.3 Use of animal and industrial wastes

Several animal and industrial wastes have been found to be very efficacious in the management of phyto-parasitic nematodes when applied to the soil. For instance steer and chicken manures reduced number of cyst and citrus nematodes and resulted in increased yields of potato and citrus (Gonzalez and Canto-Saenz, 1993). Chindo and Khan (1990) reported a significant reduction of root-knot nematode populations and root gall index following application of poultry manure on tomato (Table 3).

Hassan *et al.* (2010) reported the use of refuse dump (RD), saw dust (SD) and rice husk (RH) for nematode control with attendant increases in crop yield in tomato in northern Nigeria.

Levels of poultry manure (g/pot)	Population of *M. incognita*/kg of soil at:		Root gall index†	Root weight (g)†	Plant height at midseason (cm)†	Fruit number at harvest†	Fruit weight at harvest (g)†	Mean weight/fruit (g)‡
	Mid-season	Harvest						
7·5	1 600	8 300	5·0	54·6	34·6	6·2	235·0	37·9
15·0	600	3 100	3·1	48·2	42·4	9·0	406	45·1
30·0	540	2 800	1·0	39·1	47·8	9·6	464·0	48·3
0 (control)	12 500	17 900	6·0	62·0	31·8	5·6	202·0	36·0
C.D.	—	—	—	—	7·3	1·7	105	—

†Each value is a mean of five replicates.
‡Mean of total number of fruits.

Source; Chindo and Khan, 1990

Table 3. Effect of soil amendment with four levels of poultry manure on the development of *M. incognita* and growth of tomato cv. Enterpriser in the greenhouse.

Refuse dump was found to perform best compared to rice husk (RH) and sawdust (SD) and this was attributed to the lower C: N ratio of the RD compared to SD and RH (Table 4&5).

Treatment	Rates of application (t/ha)	Nematode populations per 500 cm^3			Nematode populations per 10 g root	Number of galls per 10 g root	Egg-masses per 10 g root	Root galls indices
		initial (Pi)	mid-season (Pm)	final (Pf)				
	15	40.0^{a}	34.3^{b}	26.0^{b}	18.0^{b}	14.3^{b}	28.0^{a}	2.3^{b}
Rice husk	30	38.5^{a}	27.3^{bc}	17.8^{bc}	9.7^{d}	8.0^{cd}	14.0^{b}	1.3^{c}
	45	37.0a	22.3^{bc}	13.2^{dc}	5.3e	3.3^{dc}	8.0^{b}	1.0^{c}
	15	38.8^{a}	31.0^{bc}	24.5^{b}	14.3^{bc}	12.0^{bc}	19.0^{ab}	2.0^{b}
Sawdust	30	32.3^{a}	23.2^{bc}	18.8^{bc}	8.3^{de}	6.3^{de}	11.7^{b}	1.0^{c}
	45	30.3^{a}	20.7^{bc}	13.0^{ef}	3.7^{ef}	2.3^{e}	5.3^{b}	1.0^{c}
	15	43.2^{a}	31.3^{bc}	20.3^{b}	11.0^{cd}	10.0^{bcd}	16.0^{ab}	2.0^{b}
Refuse dump	30	35.2^{a}	24.3^{bc}	14.7^{cd}	4.3^{ef}	3.0^{de}	6.7^{b}	1.0^{c}
	45	35.7^{a}	21.5^{bc}	8.3^{fg}	2.3^{f}	1.0^{e}	2.7^{b}	1.0^{c}
	0.01	42.5^{a}	18.5^{bc}	9.8^{ef}	5.0^{ef}	3.0^{de}	7.7^{b}	1.0^{c}
Furadan	0.032	44.5^{a}	15.3^{bc}	6.0^{g}	2.3^{f}	1.0^{e}	3.0^{b}	1.0^{c}
	0.064	36.2^{a}	11.5^{c}	2.8^{g}	1.0^{f}	1.0^{e}	1.7^{b}	1.0^{c}
Non-amended (Control)	-	37.0^{a}	53.0^{a}	66.8^{a}	45.3^{a}	39.7^{a}	38.3^{a}	4.3^{a}

Means followed by the same letter within each column are not significantly different ($P = 0.05$) as indicated by Student-Newman-Kuel's (SNK) test

Table 4. Effect of soil amendment with three organic wastes on the number of galls, egg masses and populations of *Meloidogyne* spp. on tomato in the villages of Arewaci and Kurmi Bomo of Zaria, Nigeria

This is in conformity with the report of Miller *et al.*, (1973) that availability of more nitrogen enhances the ability of the organic amendment to control nematodes. Similar achievements of nematode control through the use of several organic amendments have been reported by other workers (Abubakar and Adamu, 2004; Abubakar and Majeed, 2000; Khan and Shaukat 2002; Nwanguma and Awoderu, 2002; Nico *et al.*, 2004). The abundance of refuse dump, industrial sawdust and rice husk all over Nigeria and most developing countries makes them very suitable candidates for deployment as soil organic amendments for the management of phyto-parasitic nematodes.

Treatment	Rates of application (t/ha)	Tomato (fruit) yield (t/ha)	Plant dry weight (g)		Plant shoot height	Plant root length
			shoot	root		
Rice husk	15	4.7^{gh}	139.8^{j}	44.2^{i}	29.0^{e}	10.4^{fg}
	30	5.8^{fg}	144.4^{ij}	46.5^{gh}	29.1^{e}	11.2^{de}
	45	7.8^{d}	149.8^{hi}	48.1^{g}	33.7^{cd}	11.8^{de}
Sawdust	15	5.4^{fg}	159.7^{g}	42.4^{j}	29.9^{e}	10.3^{fg}
	30	6.6^{e}	164.4^{fg}	46.2^{hi}	31.3^{de}	10.8^{ef}
	45	8.9^{c}	171.3^{f}	47.6^{f}	36.2^{c}	11.6^{de}
Refuse dump	15	5.6^{fg}	209.4^{e}	55.4^{c}	30.0^{de}	13.1^{b}
	30	7.9^{dc}	215.6^{de}	57.9^{b}	32.7^{de}	14.2^{a}
	45	9.9^{b}	221.6^{cd}	60.3^{a}	40.9^{b}	14.8^{a}
Furadan	0.01	6.2^{ef}	227.3^{bc}	50.9^{e}	39.1^{b}	11.6^{de}
	0.032	8.0^{cd}	233.2^{ab}	53.2^{d}	42.0^{b}	12.0^{cd}
	0.064	11.8^{a}	239.7^{a}	55.5^{c}	51.5^{a}	12.6^{bc}
Non- amended (Control)	–	4.8^{gh}	127.7^{k}	45.8^{h}	28.9^{e}	9.9^{g}

Means followed by the same letter within each column are not significantly different ($P = 0.05$) as indicated by Student-New-man-Kuel's (SNK) test

Source; Hassan, *et. al.*, 2010

Table 5. Effect of soil amendment with three organic wastes on the yield and growth of tomato in the villages of Arewaci and Kurmi Bomo of Zaria, Nigeria

The beneficial effects of organic wastes are both direct and indirect. They affect the nematodes directly by releasing toxic products (after decomposition) that kill or inactivate the nematodes (Bello *et al.*, 2006). They indirectly control nematode effects by increasing soil fertility to the advantage of the crop (Boehm *et al.*, 1993). In addition to soil fertility, soil amendment with organic matter may also alter soil physical and chemical properties, and thereby affecting soil microflora (Huang and Huang, 1993).

Nematodes are important participants in the underground energy transfer system. They consume living plant material, fungi, bacteria, mites, insects and each other and are themselves consumed in turn. Some fungi do capture nematodes with traps, sticky knobs and other specialized structures (Dropkin, 1980; Jaffe *et al.*, 1998; Kumar *et al.*, 2011) (Table 6 &7).

There is substantial evidence that the addition of organic matter in the form of compost or manure will decrease nematode populations and associated damage to crops (Akhtar and Alam, 1993; Oka and Yerumiyahu, 2002; Stirling and Smith, 1991; Walker, 2004).

The fungus, *Arthrobotrys* species is a nematode-trapping fungus, which produces constricting rings for capturing and killing nematodes. The biocontrol efficiency of this fungus in reducing the population of *Meliodogyne javanica* was described by Galper *et al.* (1995). However, excellent control of root knot nematodes of vegetables following the application of granular formulations of *A. dactyloides* in pot and field was obtained by Sterling *et al.* (1998) and Sterling and Smith (1998). Hoffmann –Hergarten and Sikora (1993) also reported that the efficacy of *A. dactyloides* and some other *Arthrobotrys spp.* was enhanced with mustard as green manure and barley straw as soil amendment against early penetration of rape roots by *Heterodera schachtii.* Kumar and Singh (2005) reported that *A.*

dactyloides with cow dung manure reduced infection of plants for 10 weeks due to well developed roots that protected the initial stage of infection by capturing and killing of nematodes by this fungus. These findings are in consonance with similar reports by Sterling *et al.*, (1998), and Stirling and Smith (1998) where formulation of *A. dactyloides* caused 57 - 96% reduction in number of root-knots and 75- 80% reduction in number of nematodes per plant in tomato in pot and field experiments, respectively.

	Isolate				
	Trapping (%)				
Nematode	A	B	C	D	E
Meloidogyne incognita (J_2)	87.0 a 1	83.3 a 1	86.7 a 1	92.3 a 1	84.3 a 1
M. graminicola (J_2)	82.3 a 2	79.0 a 1	80.7 a 2	88.7 a 2	80.3 a 1
Hoplolaimus indicus	0.0 a 3	0.0 a 2	0.0 a 3	0.0 a 3	0.0 a 2
Helicotylenchus dihystera	18.7 a 4	16.3 a 3	19.0 a 4	24.7 a 4	20.7 a 3
Xiphinema basiri	0.0 a 3	0.0 a 2	0.0 a 3	0.0 a 3	0.0 a 2
Tylenchorynchus brassicae	5.3 a 5	4.0 a 4	5.7 a 5	9.0 a 5	4.7 a 5

J_2 = second-stage juvenile.
Data with same letter (a) show non-significant difference of row data among isolates by completely randomised design (CRD) test at $p \leq 0.05$.
Data with different digits (1, 2, 3, 4 and 5) show significant difference of column data among temperatures by completely randomised design (CRD) test at $p \leq 0.05$.

Source; Kumar *et.al.*, 2011

Table 6. *In vitro* trapping of plant parasitic nematodes by direct formed rings of five isolates of *Dactylaria brochopaga* after 12 h of inoculation.

Stage of development	*Helicotylenchus dihystera*	*Tylenchorynchus brassicae*	*Meloidogyne incognita*	*Meloidogyne graminicola*
Numbering of constricting ring per nematode (pre-capturing)	1–2	1–3	1–4	1–4
Inflation of ring cells	0.3 tol.5 h	0.2 to 1.3 h	10–32 min	15–45 min
Killing	2.2 to 6.3 h	2.5 to 8 h	1–4 h	1.3–5 h
Mycelial growth outside of nematode body	13–15 h	11–14 h	7–10 h	8.2–11 h
Numbering of rings formed (after mycelial growth out side nematode body)	60–110	36–64	28–52	22–43
Mycelium visible within nematode body	3–5 days	3–5 days	3–4 days	3–4 days

Source; Kumar *et.al.*, 2011

Table 7. Development of *Dactylaria brochopaga* in some important phytonematodes.

Generally, refuse dump, composts and some industrial wastes are abundant all over the major cities in the developing world. With support from governments and some private organizations, the abundance of these wastes can be channeled into our agricultural systems for the management of plant parasitic nematodes and other plant diseases.

2.4 Deployment of allelopatic plants in the management of phyto-parasitic nematodes

Production of allelopathic chemicals that function as nematode antagonistic compounds has been demonstrated in many plants such as castor bean, chrysanthemum, velvet bean, sesame, jack bean, crotalaria, sorghum-sudan, indigo, tephrosia, neem, *Tamarindus indica*, flame of the forest. These chemicals include saponins, tannins, polythienyls, glucosiniolates, cyanogenic glucosides, alkaloids, lipids, terpenoids, triterpenoids and phenolics, among others. When grown as allelopathic cover crops, bioactive compounds are exuded during the growing season or released during green manure decomposition. Sunn hemp, a typical legume, and sorghum-Sudan, a prolific grass plant grown for its biomass, are popular nematode-suppressive cover crops that produce the allelochemicals known as monocrotaline and dhurrin, respectively (Chitwood, 2002, Grossman, 1988, Hackney and Dickerson, 1975. Ball-Coelho *et al.*, 2003) found that using forage pearl millet (Canadian Hybrid 101) and marigold (rakerjack) as rotation crops with potatoes resulted in fewer root lesion nematodes and increased potato yield than rotation with rye.

2.5 Deployment of plant extracts in the management of phyto–parasitic nematodes

The use of plant extracts is one of the promising tools being investigated for the management of nematode diseases. They are relatively cheap, easy to apply, with minimal environmental hazards and have the capacity to structurally and nutritionally improve the soil health. Several parts of neem tree and their extracts are known to exhibit nematicidal activities (Bello *et al.*, 2006 ((Fig.1&2), Mojundar, 1995, Raguraman *et. al.*, 2004, Suresh *et al.*, 2004), and many neem based pesticidal formulations have been developed and marketed.

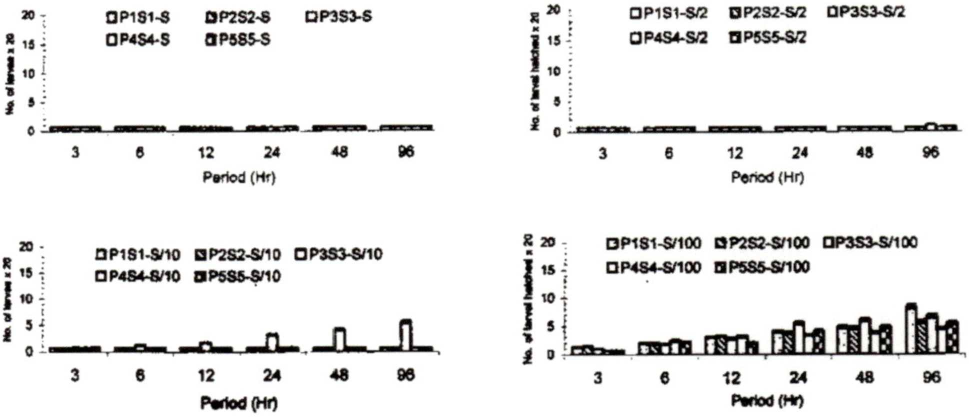

Fig. 1. Effect of water-soluble fraction of seed (S) extracts of *Tamarindus indica* (P1), *Cassia siamea* (P2), *Isoberlinia doka* (P3), *Delonix regia* (P4) and *Cassia sieberiana* (P5) on egg hatch of *Meloidogyne incognita* at different concentrations and time

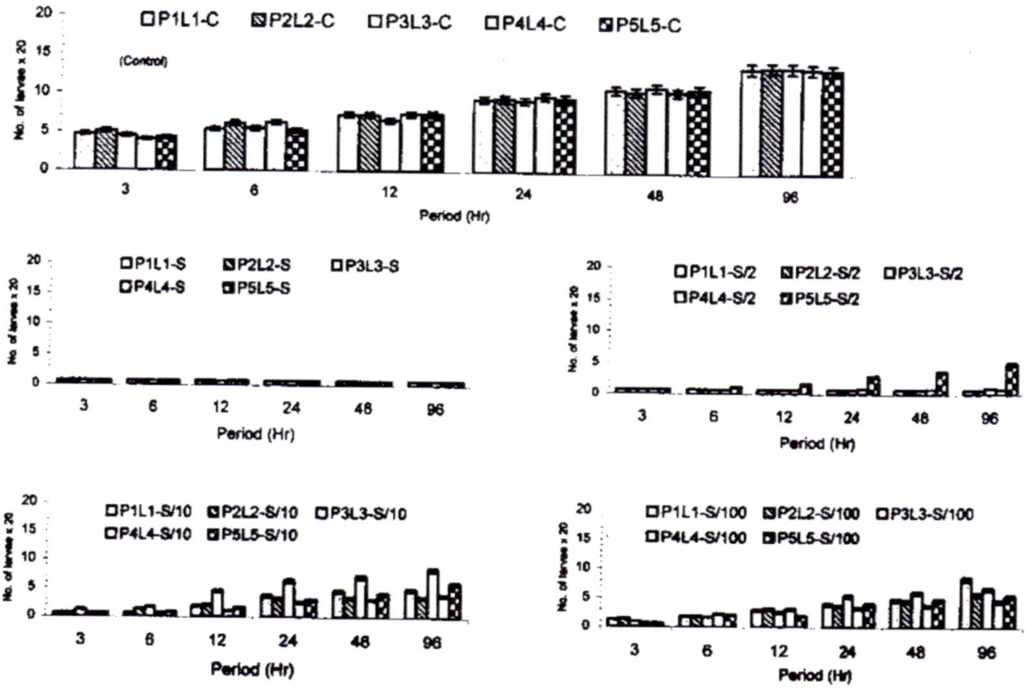

Source; Bello *et. al.*, 2006

Fig. 2. Effect of water-soluble fraction of the leaf (L) extracts of Tamarindus indica (P1), Cassia siamea (P2), Isoberlinia doka (P3), Delonix regia (P4) and Cassia sieberiana (P5) on egg hatch of Meloidognye incognita at different concentrations and time

The extracts of five organic waste of citrus (cv. Late Valencia), cocoa bean testa, rice husk, poultry manure and oil palm bunch were reported to have inhibitory effect in the egg masses of *M. incognita* (Osei *et al.*, 2011). Adegbite *and Adesiyan (*2005) reported that the root extracts of Siam weed, neem, lemon grass and castor bean were found to have nematicidal properties. Rakesh *et al.* (2000) determined the efficacy of essential oils in *Cymbopogon martinii, Mentha aivensis* and *Ociumum basilicum* on the production potential of root-knot nematode, *M. incognita* and growth of black henbane, *Hyoscyamus niger*. Reduction of population of *M.incognita* by all the oils was achieved. The effect of aqueous leaf extracts of seven plants species on hatching and mortality of second stage juneniles of *M. incognita* were tested. There was a great reduction in hatching and an increase in nematode mortality with *Murraya koenigii* (curry leaf), *Jasminum sambac* (Jasmine), *Citrus aurantifolis (*sour orange*), Zizyphus jujuba (ber), Hibiscus rarasieusis (*China rose*),* and *Justicia gandurosa (J. gendaussa*) leaf extracts (Padhi *et al.*, 2000). Bello *et al.* (2006) and Bello (2010) reported mortality of *M. incognita* following the use of water extracts *of A. indica, Tamarmdus indica, Cassia sieberiana, Cassia siamea, and Delonix regia*

Bharadwaj and Sharma (2007) observed that higher concentration levels of *Carica papaya* inhibited hatching of *M. incognita.* Similarly. The plant extract of stored pulpified peels of lemon, orange and grape fruit demonstrated significant nematicidal activity against *M. incognita* second stage juveniles after 48 hours of exposure (Tsai, 2008). Also, root exudates of *Gaillardia pulchella* were reported to be lethal to J2 of *M. incognita* and inhibitory to egg hatch at the concentration of 250 ppm or higher (Tsay, *et al.*, 2004).

Chemical based conventional systems of agricultural production have created many sources of pollution that either directly or in indirectly contribute to degradation of the environment and destruction of our natural resource - base. In this situation organic waste could be utilized both for the control of plant parasitic nematodes and other plants pathogenic diseases and improvement of soils and maintenance of a productive environment. For sustainability of agriculture in the developing economies, farmers should divorce themselves from the synthetic pesticides strategy for phyto-parasitic nematode management and marry the phytochemical option which is non-toxic to man and its environment. Most of these plants are richly available, biodegradable and affordable to the peasant farmers in the developing world.

3. Challenges and prospects in the utilization of organic wastes for the management of phyto-parasitic nematodes

The deployment of organic materials for the management of phyto-parasitic nematodes in modern day agriculture is pregnant with several challenges. These include among others initial fear of the unknown, dosage labour requirement and financial constraints

3.1 Fear of the unknown

The adoption of any new farming technology is often received by farmers with a lot of skeptism because of fear of the implications of the new technology on the productivity of their crops. Thus, adoption of such technologies is often slow until when fully convinced of its advantages over the traditional systems. Experience has shown that the transition from conventional agriculture to nature farming or organic farming can involve certain risks, such as initial lower yields and increased pest problems (James, 1994). However, once the transition period is over, which might take several years, most farmers find their new farming systems to be stable, productive, manageable, and profitable. In this case, the use of organic wastes will be beneficial through abundance of beneficial micro-organisms (characteristic of organically amended soils) which can fix atmospheric nitrogen, decompose organic wastes and residues, detoxify pesticides, suppress plant diseases and soil-borne pathogens, enhance nutrient cycling and produce bioactive compounds such as vitamins hormones and enzymes that stimulate plant growth (Higa, 1995). Besides, amendments may increase soil populations of micro-organisms antagonistic to nematodes, but are also known to release several toxic compounds during their decomposition in soil that act directly by poisoning the phyto-parasitic nematodes (Oka and Pivonia, 2002).

3.2 Dosage/Application rate

The quantities of organic wastes usually required per unit area are large.This poses problems of acquisition transportation and application particularly in large scale farms. Fortunately, in Nigeria and other developing countries, these wastes are in abundance. Large quantities of refuse dump sites, rice and other cereal straws, industrial wastes such as saw dust, rice husk, by-products of breweries, agro-processing plants etc abound. Concerted efforts by governments, organizations, non-governmental organizations (NGOs), research centers etc. are needed to mobilize these resources for use either directly or transformed into other products that can be utilized more easily by the farmers. In Taiwan for instance, fertilizers and organic wastes have been transformed into different products that are used to control plant diseases including nematodes (Huang and Huang, 1993; Huang and Kuhlman, 1991; Huang *et al.*, 2003).

3.3 Labor requirement

Traditionally organic farming is labor and knowledge - intensive whereas conventional farming is capital intensive, requiring more energy and manufactured inputs (Halberg, 2006). This, however, is not a serious drawback in most developing economies .There is abundance of idle labour which can be readily deployed to the movement and application of these wastes to work in farms thereby mitigating the myriad of social ills that is often associated with such idle minds.

3.4 Financial constraints

Research and development in organic farming is normally constrained by scarce funding from government and large commercial stakeholders, and smaller commercial players are generally unable to allocate funds for research and development. In order to have a breakthrough, research organizations such as the Colloquium of Organic Research in the United Kingdom (UK) and the Scientific Committee for Organic Agriculture Research in the USA should be formed in the developing countries such as Nigeria to boost agriculture and provide employment for the increasing population.

Organic agriculture in developing economy can be improved upon with adequate funding, removal of production subsidies that have adverse economic, social and environmental effects, investment in agricultural science and technology that can sustain the necessary increase of food supply without harmful tradeoffs involving excessive use of water, nutrients or pesticides.

4. Conclussion

In view of the foregoing, it is clear that synthetic pesticide-based conventional system of agricultural production which has created many sources of pollution either directly or indirectly, contributed to degradation of the environment and destruction of our natural resource needs to be critically examined. This is with the view to minimizing usage of these compounds and deploying much more effective, cost effective and environmentally friendly strategies that will ensure good health of our people and enhance the stability of our agricultural soils. An area that appears to hold the greatest promise for technological advances in crop production, crop protection and natural resource conservation is that of organic wastes and organic materials. The generation of solid waste has been increasing steadily after the past ten years due to rising population, urbanization and industrialization in Nigeria and most developing countries. In the early 1970s, prior to the discovery of oil in Nigeria, municipal wastes were managed as compost manure and used as organic amendments. The onset of oil wealth changed lifestyle patterns leading to increased generation of varied components of municipal solid wastes which can be channeled towards improvement in crop production.

5. References

Abubakar, U., Adamu, T. (2004). Control of *Meloidogyne incognita* (kofoid and White) Chitwood of tomato (*Lycopersicon lycopersicum* Karst) using camel dung. *Journal of Tropical Biosciences*, Vol. 47, (2004), pp.1–3. ISSN: 1007-3515

Abubakar, U., Majeed, Q. (2000). Use of animal manure for the control of root-knot nematodes of tomato. *Journal of Agriculture and Environment* Vol. 1, No. 12, (2000), pp. 29-33. ISSN 1573-322X

Adegbite, A.A and Adesiyan S.O. (2005). Root extracts of plants to control root-knot nematode on edible soybean. *World Journal of Agricultural Science* Vol.1, No.1(January, 2005), pp. 18-20, ISSN1817-3047

Akhtar, M. (1993). Utilization of plant-origin waste materials for the control of parasitic nematodes. *Bioresource Technology* Vol.46 (1993), pp. 255-257,

Akhtar,M and Alam, M. M. (1993). Control of plant-parasitic nematodes by 'Nimin'-an urea-coating agent and some Splant oils. *Zetschrift fur Planzenkrankhei ten und Pflanzenschutz* Vol. 100 (1993), pp.337-342,

Akhtar, M., and Malik, A. (2000). Role of organic soil amendments and soil organisms in the biological control of plant-parasitic nematodes. A review. *Bioresources Technology*, Vol. 74 (2000), pp. 35-47

Akhtar, M, (1998). Biological control of plant parasitic nematodes by neem products in agricultural soil. *Applied Soil Ecology*, Vol.7, (1998), pp. 219-223. ISSN 0105-8568

Andres, I. N. Rafael, M. J. and Pablo, C. (2004). Control of root-knot nematodes by composted agro-industrial wastes in potting mixtures, *Crop Protection* Vol. 23 (2004), pp. 581-587, ISSN 0306-3941

Ball-Coelho, B., Bruin, A.J., Roy, R.C and Riga, E. (2003). Forage pearl millet and marigold as rotation crops for biological control of root-lesion nematodes in potato. *Agronomy Journal*, Vol. 95, No, 2. (2003), pp. 282- 292. ISSN 0002- 1962.

Bello, L.Y., P.S Chindo.,P. S Marley, and M.D. Alegbejo. (2006). Effects of some plants extracts on larval hatch on the root-Knot nematode *Meloidogyne incognita. Archives of Phytopathology and Plant Protection, Vol.* 39 (2006), No.4, pp. 253-257. ISSN 007-2425

Bello, L.Y and M.O. Abogunde (2010). Effect of *Delonix regia* leaf extract on egg hatch and larva mortality of root- knot nematodes *Meloidogyne incognita. Journal of applied Agricultural Research,* Agricultural Research Control of Nigeria (ARIN) Vol. 2: pp. 113-117.

Bharadway A, and Sharma, S. (2007). Effects of some plant extracts on hatch of *Melodogyne incognita* eggs. *Int. J. Bot.* Vol.3 (2007), pp. 312-316, ISSN18119700

Boehm, M.J., Madden, L.V., and Hoiting, H.A.J. (1993). Effect of organic matter decomposition level on bacterial species diversity and composition in relationship to Pythium damping-off severity. *Applied Environmental Microbiology*, Vol.59, (1993), pp.4171-4179. ISSN: 0717-3458.

Chindo, P.S. and Khan, F.A. (1990). Control of root-knot nematodes (Meloidogyne spp.) on omato (*Lycopersicon esculentum* Mill.) with poultry manure. *Tropical Pest Mangement*, Vol.36, (1990), No. 4, pp. 332-335,

Chitwood, D. J. (2002). Phytochemical based strategies for nematode control. *Annual Review of Phytopathology* Vol. 40 (2002), pp. 221-249. ISSN: 1392-3196

Dropkin, V. H. (1980). Introduction to plant nemathology. John Wiley and Sons. New York. NY. pp. 38-44, 242-246, 256.

Egonjobi, O.A, Afolami, S.O., (1976). Effect of Neem (*Azadirachta indica)* leaf extract on population of *Pratylenchus brachyurus* on the growth and yield of maize. *Nematolologia*, Vol.22 (1976), pp. 125-132. ISSN 0391-9749

FAO. (1989). Sustainable agricultural production: Implications for International Agricultural Research. *Technical Advisory Committee, CGIAR, FAO Research and Technical Paper* No. 4 Rome, Italy.

Galper, S. Eden L.M, Stirling G. R, Smith, L.J. (1995). Simple Screening methods for assessing the predacious activity of nematodes trapping fungi. *Nematologica* Vol.41, (2995), pp. 130-140, ISSN 0103-8478.

Gonzalez, A., Canto-Sanenz, M. (1993). Comparism of five organic amendments for the control of *Globodera pallida* in microplots in Peru. *Nematropica, Vol.* 23 (2993), pp. 133-139, ISSN: 0717-3458

Grossman, J. (1988). Research Notes: New directions in nematodes control. *The IPM Practitioner*. February (1998). pp. 1-4.

Hackney, R.W. and Dickerson, O.J. (1975). Marigold, castor bean, and chrysanthemum as controls of *Meloidogyne incognita* and *Pratylenchus alleni, Journal of Nematology*, Vol. 7, No 1, (1975), pp. 84-90, ISSN: 1388-66-45

Halberg, N. (2006). *Global development of organic agricultural challenges and prospects*. CABI pp 297.

Hassan, M.A., Chindo, P.S., Marley, P.S, Alegbejo, M.D. (2010). Management of root-knot nematodes (*Meloidogyne spp*.) on tomato (*Lycopersicon lycopersicum*) using organic wastes in Zaria, Nigeria. *Plant Protect. Science*, Vol.46, (2010), pp.34-39, ISSN 1212-2580.

Higa, T. (1995). Effective microorganisms: their role in Kyusei nature farming and sustainable agriculture. In J.F. Parr, S.B. Hornick and M.E. Simpson (ed.) *Proceedings of the Third International Conference on Kyusei Nature Farming*. U.S. Department of Agriculture, Washington, D.C, USA. (In Press).

Hoffman- Hergartan, S, Sikora R.A. (1993). Studies on increasing the activities of nematodes trapping fungi against early attack by *Heterodera sehachtii* using organic fertilizer. *Z. Pfikr pfish*. 100:170-175.

Hoiting, H.A.J.,and Boehm, M.J. (1999). Biocontrol within the context of soil microbial communities: a substrate- dependent phenomenon, *Annual Review of Phytopathology*, Vol. 37, (1999), pp.427-446, ISSN 0021-8596.

Huang H.C. and Huang, J.W. (1993). Prospects for control of soil-borne plant pathogens by soil amendments. *Current Topics in Bot. Research, vol.* 1 (1993), pp. 223-235.

Hungalle, N., Lal, R, Terkule., C. H. H. (1986). Amelioration of physical properties by Mucuna after mechanized land clearing of tropical rain forest. *Soil Science, vol.* 141 (1986), pp. 219-224. ISSN 0973-9424.

Huang, J.W, Hsieh, T.F and Sun, S.K. (2003a). Sustainable management of soil-borne vegetable crop disease . In: *Advances in Plant Disease Management*, H.C. Huang and S.N. Acharya (Ed.), 107-119, ISSN 1230- 0462, Research Signpost, Kerala, India.

Huang, J.W. and Kuhlman, E.G. (1991). Mechanisms inhibiting damping-off pathogens of slash pine seedlings with a formulated soil amendment. *Phytopathology* Vol. 81(1991), pp.171-177. ISSN 0191-2917

Jaffe, B.A., Ferris, H. and Scow, K.M. (1998). Nematode trapping fungi in organic and conventional cropping systems. *Phytopathology*, vol.88 (1998), pp. 344-350, ISSN 1553-7374.

James, F.P. 1994. Beneficial and Effective Microorganisms. In: *Preservation of Natural Resources and the Environment*, T. Higa and F.P. James (Ed.), International Nature Farming Research Center, Afami, Japan.

Kang, B.T., Sipkens, C., Wilson, G.F., Nangju, D. (1981). Leucaena (Leuc*eaena leucocephala* (Lal) de wit) prunings as nitrogen sources for maize. (*Zea mays L*). *Fertelizer Research, Vol.* 2(1981), pp.279-287, ISSN 0564- 3295

Khan A., Shaukat, S.S. (2002): Effect of some organic amendment and carbofuran on population density of four nematodes and growth and yield parameters of rice (*Oryzae sativa*). *Pakistan Journal of Zoology*. Vol. 32 (2002), pp.145-150, ISSN 1536-2442.

Kimpinski, J, Gallant, C.E, Henry, R, Macleod, J.A, Sanderson, J.B, Sturz, A.V. (2003). Effect of compost and manure soil amendments on nematodes and on yields of potato and barley: a 7-year study. *J. Nematology*, Vol.35 No. 3 (2003), pp. 289-293, ISSN 0303-6960.

Kumar, D, Singh, K.P. and Jaiswal R.K. (2005). Effect of fertilizers and neem cake amendments in soil on spore germination of *Arthrobotrys dactyloides*. *Mycobiology*, Vol. 33 No. 4, (2005), pp. 194-199.ISSN 1229-8093

Kumar, N. (2007). Studies on predacity and biocontrol potential of *Dactylaria brochopaga*. Ph.d Thesis, Banaras Hindu University, Varanasi-India.

Kumar, N., Singh, R.K. and Singh K.P. (2011). Occurrence and colonization of nematophagous fungi in different substrates, agricultural soils and root-knots. *Archives of Phytopathology and Plant Protection, Vol.* 44 (2011)s, No.12, pp.1182-1195. ISSN 1229-8093

Kumar, N., Chindo, P.S. and Singh, K.P. (2011). The trapping fungus *Dactylaria brochopaga*: induction of trap formation, attraction, trapping and the development in some phytonematodes. *Archives of Phytopathology and Plant Protection, Vol.* 44 (2011), NO.13, pp.1322-13334.

McSorley, R.., and Gallaher, R.N. (1995a). Effect of yard waste compost on plant-parasitic nematode densities in vegetable crops. *Journal of Nematology*, Vol. 27 (1995), pp.545-549, ISSN 0919-6765.

McSorley, R. and Gallaher, R.N. (1995b). Effect of yard waste compost on nematode densities and maize yield, *Journal of Nematology*, VOl.28, 4S, (Dec.1996), 655-660.ISSN 0303-6960

Miller, P.M., Sands, D.C., Rich, S, 1973. Effects of industrial residues, wood fiber wastes and chitin on plant parasitic nematodes and some soil borne disease. *Plant Disease Reporter* 57 (1973), pp.438-442, ISSN 1450-216X

Mojumdar, V. (1995). Effect on nematodes. The neem tree, *Azardirachta indica* A. Juss. and other miscellaneous plants: Source of unique natural products for integrated pest management, industry, and other purposes. In: Schmutterer H, editor. Weinheim, Germany: VCH, Pp 129-150.

Nico, A.I., Jimenez-Diaz, R.M., Castilla, P. (2004). Control of root-knot nematodes by composed agro-industrial wastes in potting mixtures, *Crop Protection*, Vol.23 (2004), pp.581-587, ISSN 1023-1072.

Nwanguma, E.I., Awoderu J.B 2002. The relevance of poultry and pig droppings as nematode suppressants of okra and tomato in Ibadan, Southern Western Nigeria. *Nigerian Journal of Horticultural Sciences,* Vol.6(2002), pp.67-69, ISSN118-2733ss

Oka, Y. and Pivionia, S. (2003). Effect of a nitrification inhibitor on nematicidal activity of organic and inorganic ammonia-releasing compounds against the root-knot nematode *Meloidogyne javanica.* Nematology, Vol. 5, (2003), pp.505-513.

Oka, Y. and Yerumiyahu, U. (2002). Suppressive effects of composts against the root-knot nematode *Meloidogyne javanica* on tomato. *Journal of Nematology,* Vol.4, No 8, (2002), pp.891-898, ISSN 1072- 2852.

Osei, K., Addico, R, Nafeo,A., Edu-Kwarteng, A., Agyemang, A., Danso, Y., and Sackey-Asante,J. (2011). Effect of some organic wastes on hatching of *Meloidogyne incognita* eggs. *African Journal of Agricultural Research* Vol.6(10); pp. 2255-2259. ISSN 1991-637X

Padhi, N.N., Gunanidhi, B. and Behera, G. (2000). Evaluation of nematicidal potential in ten indigenous plant species against *Melodogyne incognita. India Phytopathology* Vol. 53(2000), No.1, pp.28-31, ISSN1982- 5676

Raguraman, S., Ganapathy, N. and Venkatesan, T. (2004). Neem versus entomopathogens and natural enemies of crop pests: the potential impact and strategies. In: *Neem:Today and in the new millennium* , O. Koul and S. Wahab, (Ed), 125-182, ISSN 0889-9746, Dodrecht, The Netherlands,

Rakesh, P.A.K., S. Kumar, S. and Randy, R. (2000). Efficacy of various essential oils on the management of root- knot diseases in black henbane and aromatic plants. Challenges opportunities in New Century, Contributory Papers Centennial Conference on Species and aromatic plants. Calicut, kara, India. 20th-23rd September 2000, ISSN 0971-3328

Shiau, F.L., W.C. Chung, J.W. Huang and H.C. Huang (1999). Organic amendment of commercial culture media for improving control of *Rhizoctonia* damping-off of cabbage.*Can .J. Plant Pathol.* Vol. 2(1999), pp.368- 374, ISSN 0706-0661.

Stirling G.R, Smith, L.J. (1991). Conservation and enhancement of naturally occurring antagonistics and the role of organic matter, In: *Biological control of plant parasitic nematodes, Progress, Problems and Prospects.* CAB International, Wallingford, U.K,ISSN 0717-3458.

Stirling G R, Smith L.J. (1998). Field test of formulated products containing either *Verticillin chlamydosporium* or *Arthrobotrys dactyloides* for biological control of root-knot nematodes. *Bio control,* VOl.11(1998), pp.231- 239, ISSN 0717-3458.

Stirling, G.R, Smith, L.J. Licastro, K.A., Edem, L.U. (1998). Control of root-knot nematode with formulation of nematode trapping fungus *Arthrobotrys dactyloides. Biol control,* Vol.11(1998), pp. 224-230, ISSN0717- 3458.

Sukul, N.C., (1992). Plant antagonistic to plant-parasitic nematodes, *Indian Review of Life Science,* Vol.12(1992), pp.23-52, ISSN 0002-9440

Suresh, G, Gopalakrishnan, G, Masilamani, S. (2004). Neem for plant pathogenic fungal control: the outlook in the new millennium, 183-208, In; Koul, O, Wahab, S, (Ed.), *Neem: Today and in the new millennium,* Dordrecht, ISSN 0889-9746, Dodrecht, The Netherlands.

Tsay, T.T, WUST, Lin, Y.Y. (2004). Evaluation of Asteraceae plants for control of *Meloidogyne incognita, J. Nematology,* Vol. 36(2004), pp.36-41, ISSN0022-3004

Tsai, B.Y. (2008). Effect of peels of lemon, orange and grape fruit against *Meloidogyne incognita*. *Plant Patholog Bull*, Vol.17(2004), pp.195-201, ISSN1725-5813

Walker, G..E. (2004). Effects of *Meloidogyne javanica* and organic amendments, inorganic fertizers and nematicides on carrot growth and nematodes abundance. *Nematologia Mediterranea*, Vol. Vol.32, No. 2(2004), pp.181- 188, ISSN 0718- 1620.

Wood, R.K.S. (1996). Sustainable Agriculture: the role of Plant Pathology. *Candian Journal Journal of Plant Pathology*, Vol.18 (1996), pp.141-144, ISSN 1715-2992.

9

Landfill Management and Remediation Practices in New Jersey, United States

Casey M. Ezyske and Yang Deng[1]
Department of Earth and Environmental Studies, Montclair State University, Montclair, New Jersey
USA

1. Introduction

In 2009, the United States generated 243 million tons of municipal solid waste equaling 1.97 kg per person per day. Approximately 54% or 131.9 million tons of municipal solid waste was landfilled, with a similar percentage in 2008 and 2007, which is equivalent to a net per capita landfilling rate of 1.07 kg per person per day. Municipal solid waste includes commercial waste but does not include industrial, hazardous, or construction waste (US EPA, 2010). Therefore, approximately 7.6 million additional tons of industrial wastes are disposed of in landfills in the United States each year (EPA, 2011a). In 2003, New Jersey (a state located in the Northeast of the United States) alone generated 19.8 million tons of solid waste, with 9.5 million tons sent for disposal (NJDEP, 2006).

Landfills are the ultimate disposal of waste after recovery (i.e. recycling and reuse) and combustion, and the most acceptable and used form of solid waste disposal in the United States and throughout the world due to low costs in terms of exploitation and capital costs (Renou et al, 2008). However, municipal, commercial, industrial, hazardous, and construction materials contain nonhazardous and hazardous waste such as cleaning fluids and pesticides. Hazardous waste is harmful to the health of humans and the environment, exhibiting one of the following characteristics: toxicity, reactivity, ignitability, or corrosivity (EPA, 2011b). Non-hazardous waste includes all materials thrown in the garbage, sludge from wastewater, water, and air treatment plants, and wastes discarded from industrial, commercial, community, mining, and agricultural activities (EPA, 2011a). In the early 20th century, nonhazardous and hazardous wastes were regularly burned (Hansen & Caponi, 2009) and/or placed in unlined landfills coming into direct contact and polluting the air, water, and surrounding land (Duffy, 2008). To remedy the pollution caused by landfilling, appropriate remediation options should be performed. The most common methods for the remediation of landfills include excavation to recover recyclable materials, capping to reduce leachate generation, air sparging and soil vapor extraction to capture and remediate gases, and pump-and-treat of the leachate-contaminated plume. In contrast, modern landfills minimize the amount of landfill contamination cause through liner systems, leachate collection, and caps. The government controls landfills to ensure that they are properly operated, maintained, designed, closed, and monitored (Environmental Industry Association, 2011).

---

[1] Corresponding Author

As the human population, along with the industrial, municipal, and commercial sectors, continues to grow exponentially, the amount of waste generated will significantly increase over the years (Renou et al, 2008). The number of municipal landfills and amount of waste landfilled have declined combined with an increase in recycling and composting rates over the past 40 years in the United States (EPA, 2010). However, the majority of waste is already located in landfills (Environmental Industry Association, 2011) and landfills are still the most common form of waste disposal in the United States (EPA, 2010). As of 2003, approximately 21.3 years of landfill capacity remained in the United States, and less than ten years of capacity left in New Jersey (Hansen & Caponi, 2009).

2. Background

2.1 Environmental impacts

2.1.1 Impacts of Landfills on water, land, and air

Environmental impacts from landfills, principally caused by leachate generation and gas production, include air emissions, climate change, groundwater pollution by leachate, and relevant nuisance issues (i.e. odor, litter, vectors, and dust) (Hanson & Caponi, 2009).

When landfills consisted mainly of excavated pits, the waste would come directly into contact with and contaminate the surrounding surface and groundwater. During a precipitation event, water percolates through the landfill system creating leachate, which is highly contaminated wastewater. The composition of leachate can be categorized into four main groups: dissolved organic matters (mainly volatile fatty acids or humic-like substances); inorganic macrocomponents such as calcium, magnesium, sodium, potassium, ammonium, iron, magnesium, chloride, sulfate, and hydrogen carbonate; heavy metals like cadmium, chromium, copper, lead, nickel, and zinc; and xenobiotic organic compounds such as chlorinated organics, phenols, and pesticides (Kjeldsen et al, 2002; Renou et al, 2008). The surface runoff creates gullies and erosion, washing debris, contaminants, and sediment into nearby surface water bodies (Duffy, 2008). Landfill leachate harms surface water bodies by depleting dissolved oxygen (DO) and increasing ammonia levels altering the flora and fauna of the water body (Kjedsen et al, 2002).

Air pollution is caused via two routes, the open burning of garbage and the anaerobic degradation of the organic fraction in solid waste. The open burning of garbage creates smoke, polluting the air and producing open debris. The natural, anaerobic decomposition by microorganisms transforms the waste organic fraction into methane and carbon dioxide, which are two primary greenhouse gases (Hanson & Caponi, 2009) and may kill the surrounding vegetation. The decomposition rate and amount of gas production depend heavily on the temperature and precipitation of the area (Duffy, 2008). Methane is a potent greenhouse gas that is 23 more time potent than carbon dioxide. Even though landfills are not the leading source of greenhouse gas production, they are the primary contributor to anthropogenically produced methane. (Hanson & Caponi, 2009) Volatile organic compounds (VOCs) are also released into the air directly from the products themselves such as cleaning fluids (NSWMA, n.d).

The produced gas and generated leachate from landfills must be properly collected and treated before they move offsite and further affect environmental and human health (NSWMA, n.d.) Of note, the leachate generated from the landfill bridges solid waste with the hydrosphere (particularly groundwater) and lithosphere (i.e. soil), while the landfill gases connect solid wastes to the atmosphere. Therefore, it is vital to understand that landfill engineered sites have a potential to pollute more than one of the Earth's spheres.

2.1.2 Decomposition of solid waste in landfills

Typically, solid waste within landfills undergoes four stages of decomposition: an initial aerobic phase, an anaerobic acid phase, an initial methanogenic phase, and a stable methanogenic phase. The initial aerobic phase lasts only the first couple of days as oxygen in the voids is quickly depleted without any replenishment when the waste is covered. Therefore, an aerobic biodegradation of organic fraction of solid waste solely occur during a very short period, in which carbon dioxide is produced as a product and the temperature of the waste is increased. Leachate produced during this phase comes from direct precipitation or released from the moisture content of the waste itself (Kjeldsen et al, 2002). With the depletion of oxygen, the landfills quickly become anaerobic, and aerobic microbes dominate within the landfills, allowing fermentation to take place. Therefore, in the following anaerobic acid phase, the complex organic molecules are mostly degraded to volatile fatty acids, leading to a pH decrease. The initial methanogenic phase begins when methanogenic microorganisms grow in the waste, further transforming the volatile fatty acids to methane and carbon dioxide (Renou et al, 2008). The consumption of the organic acids raises the pH of the waste. During the stable methanogenic phase, the pH continues to increase. Methane production peaks and then declines as the amount of soluble materials decreases. The remaining waste is mainly refractory, non-biodegradable compounds like humic-like substances. The overall decomposition rates can be accelerated by a high moisture content and an initial aeration of the waste (Kjeldsen et al, 2002).

During different organic waste decomposition phases, landfill leachate and landfill gases may exhibit different characteristics. When volatile organic compounds dominate in the acid phase, leachate pH is typically at 3.0-4.0, under which heavy metals, such as calcium, magnesium, iron, and manganese, largely exist in leachate. Meanwhile, a huge number of biodegradable organic compounds are present in leachate, and 5-day biochemical oxygen demand (BOD_5) and chemical oxygen demand (COD) may reach a few tens of thousands of mg/L. And the organics are highly biodegradable characterized by a high BOD_5/COD (typically > 0.6). However, with the further decomposition into methangoenic phase and subsequent reduction in the concentration of organic acids, the leachate pH is raised to a neutral range, and the leachate organic content is significantly reduced. COD may drop to a few hundreds or thousands of mg/L, and the organic compounds are refractory with a low BOD_5/COD (typically < 0.3). And the concentrations of heavy metals in leachate greatly decrease as a result of precipitation more readily occurring at a high pH. When the landfill condition transform from aerobic to anaerobic condition, sulfate may be microbiologically reduced to hydrogen sulfide, so that the sulfate level is decreased with the landfilling time. Chloride, sodium, and potassium do not show a significant change in their concentrations throughout the decomposition, thus exhibiting an inert behavior. Ammonia-nitrogen concentrations remain high during all phases of decomposition, and thought to be the largest issue in landfill management for the long term. In leachate, monoaromatic hydrocarbons (e.g., benzene, toluene, ethylbenzene, and xylenes) and halogenated hydrocarbons are the most common xenabiotic organic compounds found. They are relatively recalcitrant. The concentrations of xenabiotic organic compounds vary broadly depending on the landfill, with respect to age and restrictions of dumping hazardous waste (Kjeldsen et al, 2002). Recently, some emerging leachate contaminants, such as perfluorinated chemicals, pharmaceuticals, and engineered nanomaterials, at trace levels have been paid special attention to. However, their fates in leachate are poorly understood. For landfill gases, oxygen and nitrogen gases predominate in the initial phase because they, trapped from air, are buried together with solid waste, reflecting the composition of air. However, carbon dioxide and methane will gradually take over as products of anaerobic degradation of organic wastes. VOCs and ammonia may be present in landfill gases. Particularly, ammonia-nitrogen exists in forms of ammonium ions and dissolved ammonia gas in leachate. During methanogenic phases, leachate pH is back to neutral and even basic, and the fraction of dissolved ammonia will be increased. Therefore, the content of ammonia in landfill gases will be relatively high at these phases, and it can be quantitatively analyzed using the Henry's law that governs the distribution of dissolved ammonia gas in leachate and ammonia gas in landfill gases.

2.2 Landfill designs

Almost everything humans do creates wastes. However, waste did not become a problem until humans left the nomadic lifestyle and starting living in communities. As the world population has increase and changed from a rural agrarian society to a urban industrial society, the disposal of waste has become more concentrated. Dumping trash in the middle of cities was common practice in the United States until scientists linked human health problems to sanitary conditions in the early 1800's. In the early 20th century North America, cities began to collect garbage and either incinerated it at a landfill or home, or placed it in an unlined landfill (NSWMA, n.d.; Duffy, 2008). One of the first landfills was created in California in 1935, which consisted of a hole in the ground occasionally covered with soil (NSWMA, n.d.). Dumps were usually small and scattered affecting many areas (Duffy, 2008). Approximately 85% of U.S. sanitary landfills are unlined (Pipkin et al, 2010) and many are not covered, coming into direct contact with and polluting the air, groundwater and soil. Open dump burning was a common practice to reduce the volume of waste and increase the remaining capacity. When a landfill was closed, soil of varying thickness and slopes were placed over the waste (Duffy, 2008).

After the passage of laws and regulations that banned open burning at dumps, waste was spread into layers and regularly compacted to reduce the total volume, increase stability, and extend the life of the landfill. Modern landfills are located, operated, designed, closed, and monitored to ensure that the environment is appropriately protected (Environmental Industry Association, 2011). Newer landfills are restricted from being built in floodplains, wetlands, fault zones, and seismic impact zones unless the landfills have structural integrity and protective measures in place to protect human and environmental health. Protective operational procedures include rejecting hazardous and bulk materials, non- containerized liquids, the restriction of open burning, securing site access, and keeping up-to-date records on groundwater, surface water, and air monitoring results. Landfills are now designed with leachate collection and liner systems to prevent the migration of leachate off-site. A liner of low permeability materials such as clay, geotextiles, or plastic, with a leachate collection and recovery system placed on top of the liner. The leachate collected are either treated on or off-site at a wastewater treatment plant, while the gases produced are burned or converted into energy (i.e. electricity, heat, steam, replacement of natural gas, or vehicle fuel). Waste is layered above the leachate collection system, compacted, and covered daily to reduce odors, vectors, fires, and blowing litter. When the landfill reaches a permitted capacity and then is closed, a final cap is placed on the top of the landfill to prevent precipitation seeping through the waste. The final cap consists of a low permeability material such as clay or synthetic material (NSWMA, n.d.). Storm water channels are constructed on and around the landfill to direct rainwater to retention ponds for erosion control and reduce surface water contamination. Lastly, a long-term monitoring plan is implemented to ensure the liner and gas/leachate collection systems are operating properly, and the surrounding or underlying groundwater is not contaminated (Environmental Industry Association, 2011). Properly designed landfills can be inexpensive means of disposal (Hanson & Caponi, 2009), but many landfills are older, poorly designed and not managed, thus causing numerous environmental impacts (NJDEP, 2006).

3. Regulations

The Solid Waste Disposal Act of 1965 was the first regulation on waste disposal in the United States, and formed the national office of solid waste. Within the following 10 years, every state had regulations on the management of solid waste, varying from the banning of open burning to requiring permits and regulations on design and operational standards (NSWMA, n.d.).

The Resource Conservation and Recovery Act (RCRA), passed by Congress in 1976, and the RCRA Hazardous and Solid Waste Amendments in 1984 granted the US Environmental Protection Agency regulatory control over the disposal of waste (Hanson & Caponi, 2009). The program was implemented to assess the problems associated with an increasing amount of municipal and industrial wastes that the nation was confronted with. RCRA separated hazardous and non-hazardous waste and mandated the Environmental Protection Agency to create design, operational, locational, environmental monitoring standards, to close or upgrade existing landfills, and secure funding for long-term assessment of the landfill (NSWMA, n.d.).

The solid waste program, under Subtitle D, requires states to create management plans, set criteria for solid waste, and restrict the use of open dumping. Subtitle D's regulations lead to the creation of larger, regional landfills and waste management companies, which improves environmental and economical integrity relative to the small, scattered dumps of the past. Larger waste management facilities are more cost effective in terms of capacity, volume, and operational resources (i.e. staff and equipment) to meet the increasing volume of waste (Duffy, 2008).

The Resource Conservation and Recovery Act addresses only active and future landfill sites, while the Comprehensive Environmental Response, Compensation, and Liability Act (CERCLA), otherwise known as Superfund, focuses on abandoned or historical sites (EPA, 2011). The Environmental Protection Agency, through the Superfund program, holds the parties responsible for clean up or if no responsible party can be identified, the Agency uses money from a special trust fund. This program is a complex, long-term cleanup process involving assessment, placement on the National Priorities List (NPL), and implementation of appropriate cleanup plans (EPA, 2011). The National Priority List is a list of the sites contaminated by hazardous waste and pollutants in the United States, eligible for long-term remedial action financed under the federal Superfund program, and guides the Environmental Protection Agency to which sites need further environmental assessments (EPA, 2011).

4. New Jersey landfills

Although the area of New Jersey ranks No. 47 in the 50 states of the United States, New Jersey is the most densely populated (462/km^2) with a population of approximately 8.4 million residents This state is faced with an increasing trend in volume of waste generation, combined with a declining trend in recycling rates, and a scarcity of open spaces to site new landfills. Compounding the problem is the large quantity of legal uncertainty regarding the permissible regulation of solid waste collection and disposal, and a marketplace that makes identifying additional disposal capacity difficult (NJDEP, 2006).

For the past thirty years, the Solid Waste Management Act has guided New Jersey in terms of the collection, transportation, and disposal of solid waste. The development of facility siting and recycling plans are the responsibility of twenty-one counties and the New Jersey Meadowlands District, and each municipality ensures the collection and disposal of solid waste adhere to the county plan (NJDEP, 2006).

In 2006, the Statewide Solid Waste Management Plan was updated from the 1993 version. Since 1993, New Jersey has undergone significant changes in terms of solid waste management including declining recycling rates, the loss of a variety of funding sources due to numerous taxes, invalidation of waste flow rules by the Federal Court, the partial deregulation of solid waste utility industry, and the state adopted the federal hazardous waste program. Two Federal Court decisions, "Atlantic Coast" and "Carbone", left many once financially secure disposal facilities with significant debt. After "Atlantic Coast" and deregulation of state control on regulatory flow, several counties controlled their waste and initiated an intra-state flow plans allowing waste to leave the state, but if the waste remains in New Jersey, it is sent to a facility in that county. Due to these changes, the resources needed to plan and execute an environmentally protective solid waste management program are not available (NJDEP, 2006).

In the mid 1970's, as old dumps were being closed and the generation of waste increased, the formation of environmentally friendly landfills could not maintain the increased waste, resulting in New Jersey becoming a net exporter of waste to neighboring states. Therefore, the state embarked on a mission to increase recycling rates while creating environmentally sound landfills for the remainder of the waste (NJDEP, 2006).

Some counties choose to create facilities using funds from revenue bonds backed by the guaranteed flow of waste to the publicly owned facility. By 1990, thirteen new facilities were built creating billions of dollars of public debt. However, a Federal Court ruling in "Atlantic Coast" invalidated this waste flow system. The public funded facilities could not modify their systems as easily as the counties that contracted with private entities and still pay for the acquired debt. These facilities have higher rates due to several aspects: the scarcity of open spaces in such a densely populated state, having to accept even the unprofitable segment of the waste, the numerous taxes and surcharges supporting recycling programs, and the need for the proper closure of landfills in the future. In certain counties, the state decided to subsidize the debt payments and cleared certain loans related to solid waste management (NJDEP, 2006).

4.1 County plans

The Statewide Waste Management Act amended in 1975 mandated districts to establish solid waste management systems with emphasis on resource recovery such as recycling, composting, and incineration to minimize the disposal of waste in landfills. In the beginning of the 1980's, New Jersey Department of Environmental Protection (NJDEP) permitted the solid waste management plans for the 22 solid waste management districts, which include the 21 counties in New Jersey and the New Jersey Meadowland Commission. Currently, New Jersey contains 16 operating landfills, five of which have resource recovery facilities (NJDEP, 2006).

+The districts/ counties use four waste management systems, including non-discriminatory bidding flow control, intrastate flow control, market participant, and free market controls. The non-discriminatory bidding flow control is brought about due to the non-discriminatory bidding process, opening the bidding of contracts to companies both in-state and out-of-state for the disposal of a county's waste. The intrastate flow control system requires that all waste should be disposed of within the same county as it was generated, unless transported out-of-state for disposal. In a market participant system, a county owned facility is permitted to compete with in and out- of- state disposal facilitates, and the free market system permits the ability to make freely agreed upon terms between the district/county, transporter, and disposal facility. Eight districts have the non-discriminatory bidding flow control, while the other districts utilize either a market participant or free market approach for disposal of the solid waste generated within their borders. (NJDEP, 2006)

4.2 Waste generation

Figure 1 depicts the solid waste disposal trends in New Jersey from 1985 to 2003 including in state and out-of-state disposal statistics. These figures illustrate a steady rise in solid waste generation during this period. This increase may be attributed to a strong economic landscape in New Jersey or a population rise.

Figure 2 shows the amounts of solid waste exported to the various neighboring states from 1990 to 2003. The export rates steadily increase for Pennsylvania and Ohio and more recently Delaware. The figure clearly shows that Pennsylvania receives the majority of New Jersey waste if it is exported out-of-state (NJDEP, 2006).

In 2003, New Jersey generated more than 19.8 million tons of solid waste, with 9.5 million tons sent for disposal. Of the 9.5 million tons disposed, sixty percent of the waste was disposed at facilities, including recycling facilities, in New Jersey, while forty percent or 3.9 million tons were sent to out-of-state facilities. The amount of exported waste has been increasing over the years (NJDEP, 2006).

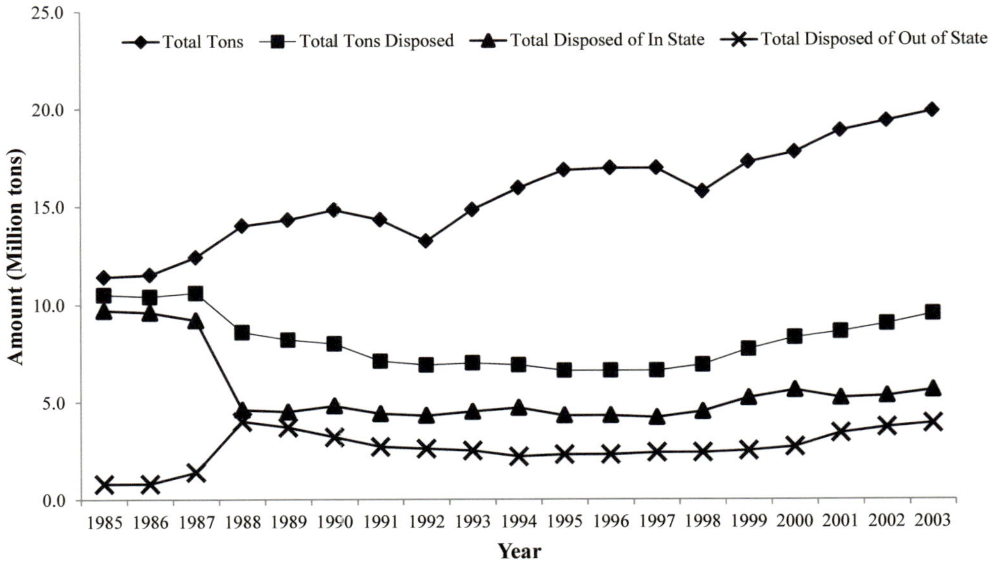

Fig. 1. New Jersey Solid Waste Trend Analysis (NJDEP, 2006)

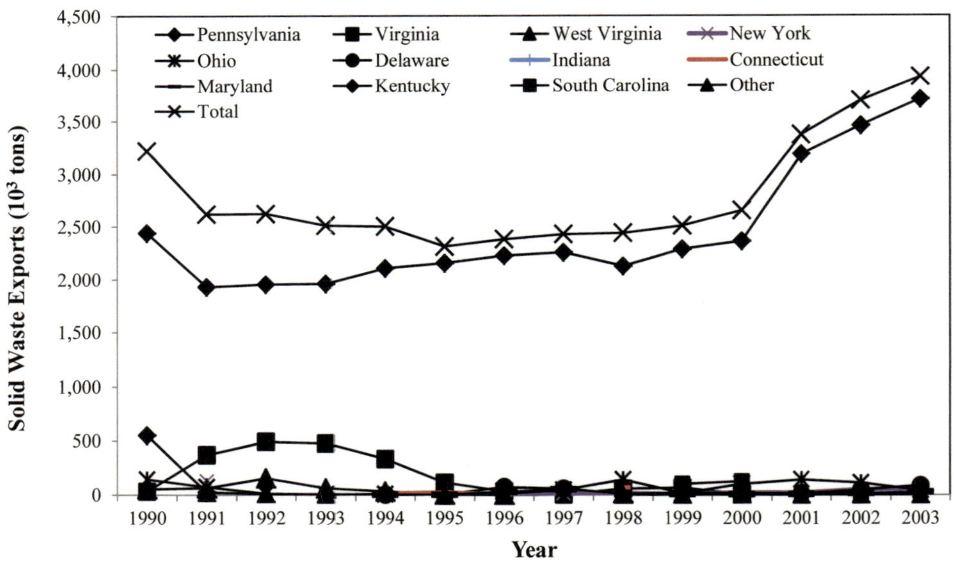

*Note: Data for 1990 through 2003 was developed from information received from solid waste transfer stations and transporter monthly reports submitted to the New Jersey Department of Environmental Protection

Fig. 2. Solid Waste Exports from New Jersey to neighboring states (in 000's of tons) (NJDEP, 2006)

The New Jersey chapter of the Solid Waste Association of North America states that about 3.6 million cubic meters of waste was disposed of in 2004 and there is a sufficient permitted capacity, 31.9 million cubic yards, remaining for the short term. This means that there is less than 10 years of landfill capacity left in New Jersey (NJDEP, 2006).

The "self sufficiency" policy of creating and preserving in state facilities that are environmentally protective and cost efficient for in-state generators has been limited by constitutional failures. Since new landfills in New Jersey are difficult to site and additional capacity at existing facilities are limited, this plan encourages activities for a sustainable landfill including leachate recirculation, use of alternative covers, and landfill mining. New Jersey will continue to identify and properly close all landfills, use public funds to remediate environmental problems, and promote brownfield redevelopment of closed landfills (NJDEP, 2006).

5. Analysis of contamination by NJ landfills

The contamination caused by active, inactive, and closed landfills in New Jersey, particularly landfill Superfund sites, is reviewed and analyzed. All the data on the landfills were acquired from US EPA, and then input into a database with regards to geographical location, contaminant type, pollution media, current status, and remediation method.

Since there were no regulatory requirements or mandatory registration for solid waste landfilling activities until the 1970s, many New Jersey landfills were poorly sited, designed, and controlled. In addition, solid waste from neighboring states was sent to New Jersey in an uncontrolled manner. The solid waste was dumped with little or no provision for cover to prevent odor, to control birds, insects, and rodents, or to minimize long-term environmental impacts. Even though New Jersey has the strictest design and performance standards for new landfills in the nation, there are many old landfills throughout New Jersey. The legacy of past landfills not designed with stringent controls for environmental protection or closed properly remains a significant challenge facing the state (NJDEP, 2006).

Most landfills established before to the mid- 1970's lacked any leachate collection or control system, discharging the leachate directly to surface and groundwater causing serious water quality impairments. And closed landfills that do not have leachate collections systems require a costly retrofitting of a system to control discharges to surface and groundwater. Landfills, operated before the relevant environmental laws were enacted, accepted all types of waste, including industrial and commercial waste. Even after the laws were enacted, commercial and industrial waste continued to be illegally dumped at many municipal landfills. Therefore, many landfills may contain a variety of hazardous wastes. Nonetheless, municipal waste contains trace amounts of different household hazardous materials as homeowners dispose of paints, cleaning agents, solvents, and pesticides. As these hazardous materials accumulate in a landfill, a significant level of hazardous substances may result (NJDEP, 2006).

The largest anthropogenic source of methane gas emissions in New Jersey is landfills, accounting for 72% or 13.3 million tons of methane emissions. Approximately 35% or 1.9 million tons of methane emissions is released from only forty-seven landfills, both open and closed. These sites should use energy recovery systems to capture and use the greenhouse gas as a renewable resource. Additional revenue is obtained when the methane gas is resold or used to generate electricity (NJDEP, 2006).

New Jersey implements energy recovery facilities at several large landfills. At one landfill, revenue is generated from the electricity sales and the carbon dioxide emission credits. All suitable landfills in New Jersey, both large and small, should develop these energy-to-recovery systems and assist in the funding to properly close the landfill and monitor gas emissions after closure (NJDEP, 2006).

In New Jersey, even if landfills do not receive wastes, they are not technically closed due to financial issues since final closure is expensive. All closure plans involve some degree of grading, landscaping, re-vegetation, site securing, drainage control, capping, and groundwater monitoring. Based upon historical experience in the solid and hazardous waste management program of the NJDEP, the following financial estimates are made. For a facility that requires the most limiting measures of closure may costs of up to $180,000 per acre, while a more detailed closure involving an impermeable cap with a single synthetic geo-membrane could cost up to $225,000 per acre. Finally, a full capping scenario of a remediation case, where substantial contamination has been identified and a 24-inch clay cap and synthetic membrane was used, the cost of closure increased up to $ 700,000 per acre (NJDEP, 2006).

The NJDEP has more than 400 registered landfills and 200 additional sites suspected not to be registered. Before January 1, 1982, landfills were not required to submit detailed closure and post closure care plans. Therefore, out of the 400 registered landfills, 166 operated after 1982 submitting detailed plans under the "Sanitary Landfill Facility Closure and Contingency Fund Act" (Closure Act), N.J.S.A. 13:1E-100. The Closure Act places regulatory control upon closure and collects taxes on the landfills, which is reserved for final site plans (NJDEP, 2006).

Among the 146 NJ Superfund sites, namely NPL sites, 45 were, at least partially, contaminated by municipal and industrial landfill activities, of which only 10 have been completely remediated. The polluted media, in terms of occurrence frequency, are groundwater (91%), soil (62%), surface water (31%), wetlands (16%), and air (11%). A breakdown of the primary contaminated media for landfills on the Superfund list in New Jersey is shown in Figure 3. Moreover, 10s-100s of thousands of people reside within 5 km from these sites and they are located nearby natural water and public parks. Particularly, the Ringwood Mines Landfill is near a major drinking water source supplied for approximately 2 million people. Contamination of drinking water sources has been occurring but to date has been offset by improvements in detection and water treatment systems. Of the 45 NJ landfill Superfund sites, the most frequently found contaminants are volatile organic compounds such as benzene, contaminating 84% of the sites, and heavy metals like lead are found in 80% of the cases. Figure 4 shows the primary contaminants of landfills on the Superfund list in New Jersey.

Landfill leachate from the most landfills contain the common contaminants at different levels such as biochemical oxygen demand (BOD), chemical oxygen demand (COD),

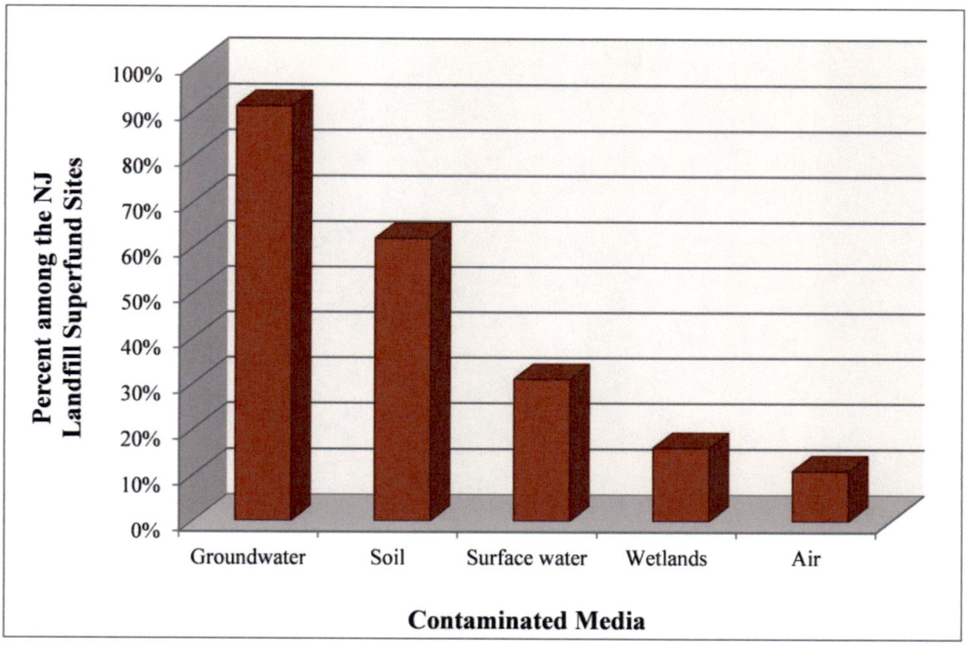

Fig. 3. Percentages of New Jersey landfill superfund sites in terms of different contaminated media

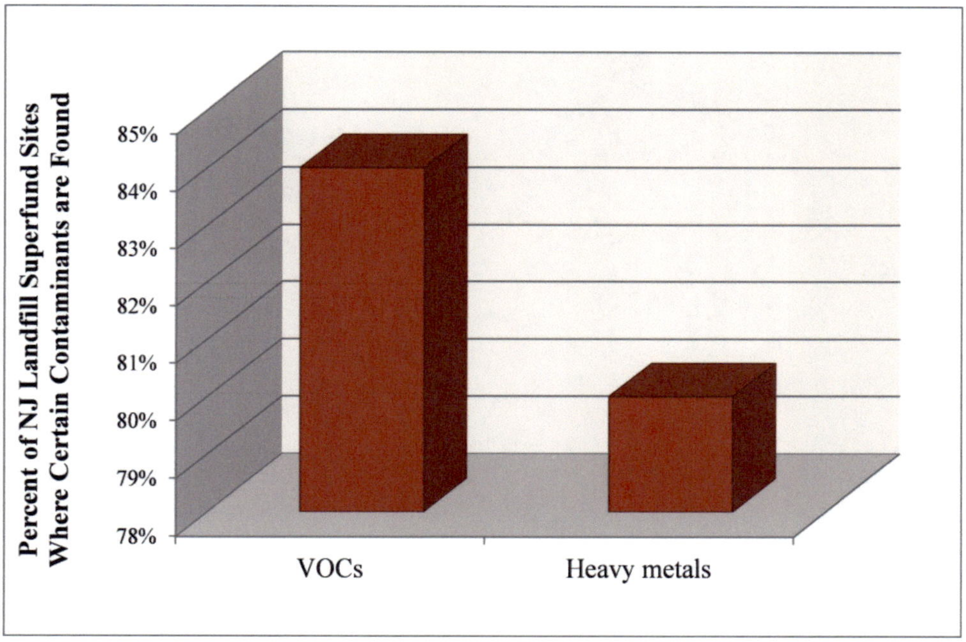

Fig. 4. Percetages of NJ landfill superfund sites where certain contaminants are found.

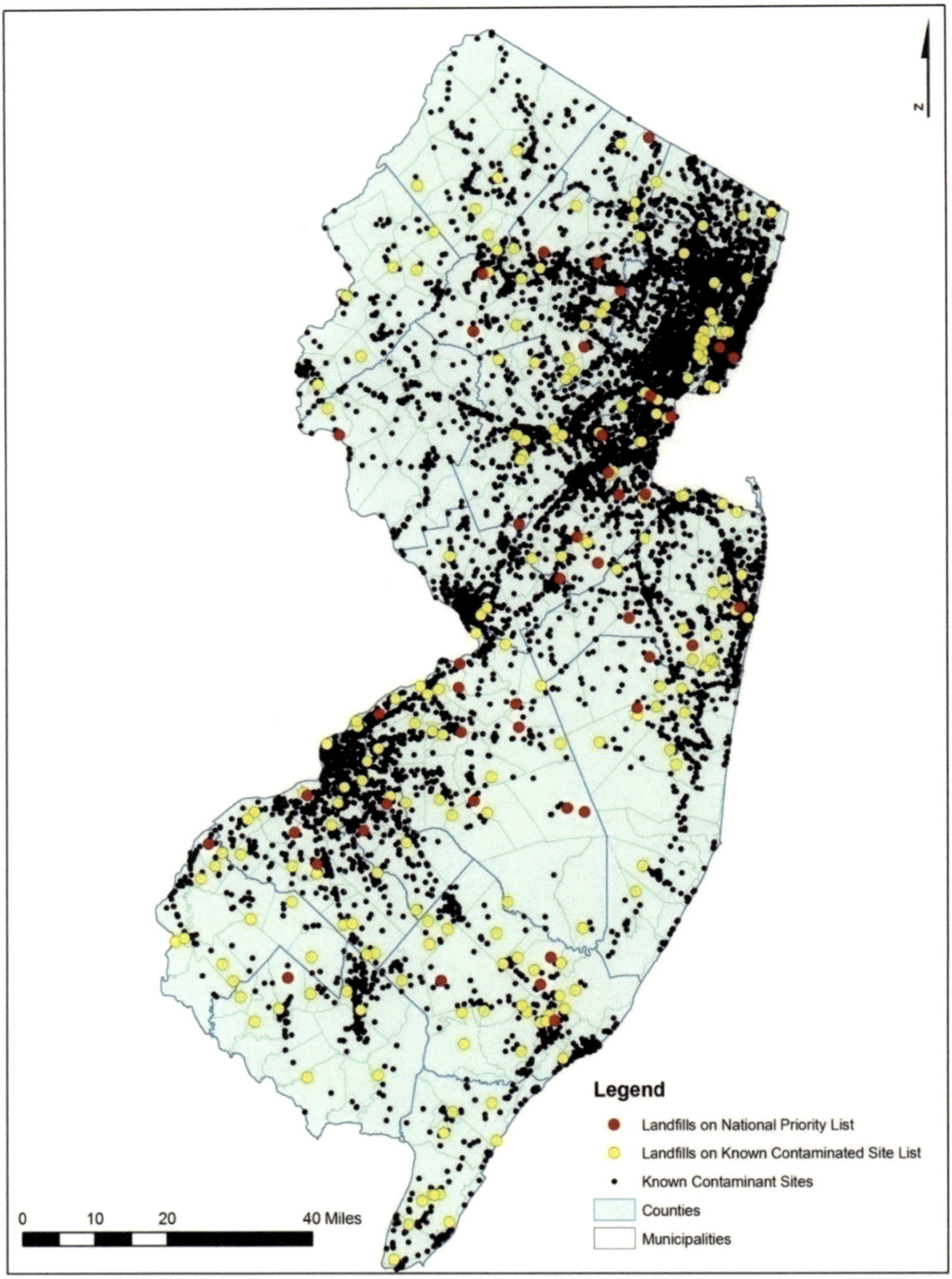

Fig. 5. New Jersey Landfills listed on the Known Contaminated Site List and the National Priority List

ammonia, heavy metals, and chlorinated hydrocarbons. However, due to old landfills accepting all types of waste and illegal dumping, the leachate from these landfills may contain uncommon pollutants (e.g. PCBs) derived from hazardous substances such astar, paint sludge, waste oils, drummed industrial waste, and medical waste. These old landfills with mixed solid wastes are usually the ones considered for redevelopment, which poses many problems with remediation (Wiley and Assadi, 2002).

New Jersey landfills listed on the Known Contaminated Site List and the National Priority List are shown in Figure 5. It would be noted that the known contaminated sites include all sites on the National Priorities List (Superfund sites) and other contaminated sites (e.g. brownfields). The majority of the landfills are concentrated the areas of New Jersey close to New York City and Philadelphia area, with a band connecting the two areas in central New Jersey. The counties with the most landfills are Atlantic, Middelsex, and Morris, with 9, 7, and 6 landfills, respectively, most likely as a result of high population density, urbanization, and industrialization.

6. Remediation methods

The frequently used remediation methods for landfills on the National Priority List in New Jersey include excavation to recover recyclable material, capping to reduce leachate generation, air sparging and soil vapor extraction to capture and remediate gases, and pump-and-treat the leachate plume .

Since siting new landfills is a lengthy, expensive endeavor and the community raises opposition to landfills nearby, New Jersey may not have suitable areas for a new landfill. Additionally, existing, operating landfills cannot adequately expand new cells. Therefore, the Statewide Waste Management Plan promotes the use of "sustainable landfills" implementing innovative technologies to extend the lifetime of the landfills. In addition to the methods mentioned above, New Jersey is researching alternative daily covers, deterring bulky wastes, landfill surcharging, and redevelopment opportunities (NJDEP, 2006).

6.1 Excavation/ landfill mining

Landfill mining consists of excavating and the subsequent processing of landfilled wasted. This procedure recovers recycled materials, cover soil, and a combustible fraction to free landfill space. The excavation techniques for landfill mining have not changed since the 1950's and resembles surface mining. The excavated mass is processed through a series of screens for sorting. The amount recovered depends heavily on the physical and chemical properties of the waste, types of mining technologies used, and the efficiencies of the applied technologies

However, if the separated materials are contaminated or have a poor quality, the viability of recovering recyclable items from old landfills in New Jersey is reduced. Landfill mining would be most advantageous when the waste is fully decomposed and stabilized such as after either aerobic or anaerobic bioreactor (NJDEP, 2006).

6.2 Landfill caps

Capping a landfill involves three layers: an upper vegetative (topsoil) layer, a drainage layer, and a low permeability layer made of a synthetic material covering two feet of compacted clay. Capping has a life span of about 50-100 years, but the performance of the cap depends on the site's environmental conditions. Cracking and erosion of caps can occur due to fluctuations in air temperatures and precipitation, and if the site is prone to subsidence and earthquakes. The cap must be adequately thick to prevent frost, and accommodate vegetative roots and burrowing animals (Vasudevan et al, 2003).

The use of temporary caps, instead of a final cap, on a filled landfill cell increases landfill space because the feet of soil typically used in a final cap is replaced with a synthetic membrane held down by removable items, like tires rather than soil. Temporary caps may be used in conjunction with leachate recirculation and active gas extraction. They are readily removable, and do not occupy much space like soil when the landfill is reopened for future landfilling activities. Some landfills are using temporary tarps to cover the waste instead of daily soil covers, increasing landfill capacity. Soil-like materials, like spray foam, can substitute soils as daily or intermediate cover material frees landfill space (NJDEP, 2006).

6.3 Landfill surcharging

When a landfill reaches final elevation levels, landfill surcharging may be implemented. The surcharging of a landfill involves the placement of a large amount of weight on top of the landfill for 6-12 months. The added weight to the top of the landfill causes enhanced settlement of the waste and increased capacity, which is recognized after the surcharge material is taken away. Clean soil is usually used as the surcharge material, which may be used elsewhere in the landfill after the surcharging process is completed (NJDEP, 2006).

6.4 Soil vapor extraction and air sparging

Soil vapor extraction (SVE) and air sparging are in-situ remediation techniques to remove vapors from polluted soil and plume, respectively (Vasudevan et al, 2003). Usually, SVE and air sparging are concurrently used in a site (EPA, 2001).Solvents, fuels (EPA, 2001), and volatile organic compounds (Vasudevan et al, 2003) are readily removed through these methods. Two types of wells are installed around the landfill, extraction wells and air injection wells. An extraction well creates a vacuum to draw the vapors to the surface, while an air injection well pumps air into the ground. The air injected stimulates the growth of aerobic microbes to enhance microbial decomposition. If the injected air is heated, the evaporation of the chemicals is accelerated.

SVE and air sparging are safe but may take years to reach full remediation depending on the size and depth of pollution, type of soil, and concentration of chemicals in the soil and groundwater. However, these methods are quicker than just relying on natural processes (EPA, 2001).

6.5 Co-treatment of landfill leachate with sewage in a wastewater treatment plant

Leachate management involves discharging to a wastewater treatment plant, pre-treatment before discharging to a wastewater treatment plant, or treatment onsite and following discharging to a nearby stream. The connection with a nearby sewer line is the most common practice in the United States. Most of municipal wastewater treatment plants use aerobic biological treatment (e.g. activated sludge process), and were specially designed to aim at biodegradable organic matters and suspended solids in sewage. Therefore, refractory organics and emerging pollutants in leachate may be poorly removed. Although much large sewage, relative to leachate, can dilute the persistent pollutants from leachate, it should be noted that these leachate pollutants are not truly removed or eliminated. Moreover, toxic chemicals in leachate (e.g. ammonia and heavy metals) may disturb microbial activities and cause unusual operation in wastewater treatment plants. In addition, sewer lines may be unavailable, be of insufficient capacity, or be disallowed for some reasons for connection to nearby treatment plants. (Spengel and Dzombak, 1991).

6.5.1 Bioreactor landfills

The recirculation of leachate back into filled cells is an essential step in a bioreactor landfill, in which microbial activities are intentionally enhanced. The recirculation of leachate provide moisture and/or oxygen to stimulate the microbial degradation of solid wastes and simultaneously reduce the amount of leachate needed for treatment. A bioreactor landfill may be either aerobic or anaerobic, to reclaim landfill space. Aerobic bioreactor landfills inject both air and leachate into the waste, while anaerobic bioreactor landfills only inject leachate into the waste. The aerobic bioreactor increases microbial digestion rates, thereby resulting in quicker settlement of the waste compared to anaerobic bioreactor landfills. In contrast, anaerobic bioreactor landfills generate more methane gas. Thus, it is a promising candidates for energy recovery projects. However, the recirculation is not commonly accepted practice among the waste management community or is not favored by regulations (NJDEP, 2006).

6.6 Deterring bulky waste

Many landfills discourage the acceptance of bulky waste since bulky waste results in large voided spaces in a landfill due to the inert, or inability, to decompose in a landfill, which is an inefficient way to use a landfill especially when this material could be recycled. Bulky waste such as tree parts, construction and demolition debris, tires, carpets, are deterred from a landfill by implementing higher fees for this material or to build recycling and recovery facilities at the landfill to reduce the amount of material landfilled. Landfills can use crushed tires or construction and demolition debris for alternatives to covers or stone (NJDEP, 2006).

6.7 Brownfield redevelopment

Brownfield redevelopment remediates and preserves existing contaminated sites, like old landfills, for use in the future. Brownfield redevelopment provides economic development by establishing new areas for businesses and industry to expand, and gives people the opportunity to gather, visit, shop, recreate, or work in different places. Brownfield redevelopment not only provides economic advantages but also brings communities together in New Jersey. However, the redevelopment of landfills is a challenge due to a variety of contaminants involved and the geological issues of building above a landfill (Wiley and Assadi, 2002). The issues associated with landfill redevelopment projects include: the size of the landfill, contaminants' types, the size and depth of plume, type and depth of waste, avoidance of open water areas, utilization of recyclable material for remediation and development, land value, finding a willing developer, regulatory guidelines, engineering designs, and financial incentives (NJDEP, 2006).

Depending on the end use for the site and landfill conditions, some sites may just need a traditional final cap and clean fill over the waste, while other sites may need to move and consolidate the waste into a more appropriate, controlled location. Some landfill redevelopments use residual materials such as contaminated sludge and recyclables to re-contour the site and surcharge the waste, which is cheaper than using several meter-deep clean fill soil. All brownfield redevelopment projects must acquire all needed permits by multiple layers of government; conduct remedial investigations of the degree of contamination, gas, and leachate contamination; investigate natural constraints such as wetlands and discharges into surface water; study public and environmental health and safety; and identify the stability and serviceability of the development structures (NJDEP, 2006).

The larger the site, the more the redevelopment project is going to cost due to the probability of more natural constraints and illegal dumping of hazardous waste, thus increasing the costs of remediation. The remedial cost per acre reaches a plateau at 130 acres or more (NJDEP, 2006).The NJDEP supports private developer's landfill closures and third party landfill closure projects. The 1996 Gormely Bill offers up to 75% in state tax credits for remediation costs, and other financial and legal incentives are provided under the 1998 Brownfield Law (NJDEP, 2006).

There are several examples of successful brownfield redevelopments projects in New Jersey. One of the largest redevelopment projects is the EnCap Golf Holdings, LLC, where several closed landfills were capped and remediated for the construction of a golf course, commercial development, and residential areas in Bergen County. Another example of a brownfield redevelopment of old landfills is the Borgata Casino on the Atlantic City Landfill (NJDEP, 2006).

7. Conclusion

In this study, the data review and analysis show that amount of municipal, industrial, and commercial waste are continuing to grow, continuously shrinking the remaining landfill space in New Jersey. Landfilling waste remains the best option for disposal, but New Jersey is a densely populated state without much capacity or areas to expand and create landfills. Even though landfill designs have improved significantly, many old landfills continue to pollute the air, groundwater, surface water, and soil. Cost benefit analysis followed by an appropriate cleanup strategy should be carefully implemented to clean up each contaminated site. While New Jersey implements innovative technologies to recover landfill space and remediate contaminated sites for redevelopment opportunities, most of these techniques continue to require many years of execution

8. Acknowledgment

This study was partially supported via the new faculty start-fund by the College of Science and Mathematics (CSAM) at Montclair State University, New Jersey, USA.

9. References

Environmental Protection Agency (2001). A Citizen's Guide to Soil Vapor Extraction and Air Sparging. *Office of Solid Waste and Emergency Response.* Obtained July 2011 from: http://www.clu-in.org/download/citizens/citsve.pdf

Duffy, D.P. (2008). MSW Processing Past, Present, and Future, The Journal for Municipal Solid Waste Professionals, Retrieved from: www.mswmanagement.com/elements-2008/msw-processing-past.aspx

Environmental Industry Association. (2011). Municipal Solid Waste Landfills, July 2011. Available from: http://www.environmentalistseveryday.org/issues-solid-waste-technologies-regulations/landfills-garbage-disposal/index.php

Environmental Protection Agency. (2010) Municipal Solid Waste in the United States: 2009 Facts and Figures. *Office of Solid Waste.* Obtained July 2011 from: http://epa.gov/osw/nonhaz/municipal/pubs/msw2009rpt.pdf

Environmental Protection Agency (2011a). Wastes- Non-Hazardous Waste. Obtained June 2011 from: www.epa.gov/wastes/nonhaz/

Environmental Protection Agency (2011b) Wastes- Hazardous Waste. Obtained June 2011 from: www.epa.gov/wastes/hazard/

Hanson, D.L. & F.R. Caponi. (2009). *US Landfill Disposal the Big Picture,* The Journal for Municipal Waste Professional, Retrieved from: www.mswmanagement.com/elements-2009/us-landfill-disposal.aspx

Kjeldsen, P., M.A. Barlaz, A.P. Rooker, A. Baun, A. Ledin, & T.H. Christensen. (2002) Present and Long-term Composition of MSW Landfill Leachate: A Review. *Environmental Science and Technology.* 32,4 pp. 297-336

National Solid Wastes Management Association (n.d.) Modern Landfills: A Cry from the Past. Obtained June 2011 from: http://www.environmentalistseveryday.org/docs/research-bulletin/Research-Bulletin-Modern-Landfill.pdf

New Jersey Department of Environmental Protection. (2006) *Solid Waste Management and Sludge Management State Plan Update.* Solid and Hazardous Waste Management Program.

Pipkin, Trent, Hazlett, and Bierman. Geology and the Environment (6 ed.), Brooks/Cole, 2010

Renou S., J.G. Givaudan, S. Poulain, F. Dirassouyan, and P. Moulin. (2008). Landfill leachate treatment: review and opportunity. Journal of Hazardous Materials. 150: pp. 468-493

Spengel D.B. & D.A. Dzombak. (1991). Treatment of landfill leachate with rotating biological contractors: bench-scale experiments. *Research Journal of the Water Pollution Control Federation*: 63(7) pp. 971

Vasudevan N.K., S. Vedachalam, & D. Sridhar. (2003). Study of the Various Methods of Landfill Remediation. Workshop on Sustainable Landfill Management. pp. 309-315

Wiley, J.B. & Assadi B. (2002) Redevelopment Potential of Landfills: A Case Study of Six New Jersey Projects, Federation of New York Solid Waste Associations Solid Waste/ Recycling Conference, Lake George, NY

10

Synergisms between Compost and Biochar for Sustainable Soil Amelioration

Daniel Fischer* and Bruno Glaser
Martin-Luther-University Halle-Wittenberg, Institute of Agricultural and Nutritional Sciences, Soil Biogeochemistry, Halle, Germany

1. Introduction

Driven by climate change and population growth, increasing human pressure on land forces conversion of natural landscapes to agricultural fields and pastures while simultaneously depleting land currently under agricultural use (Lal, 2009). Consequently, a vicious circle develops; further aggravating climate change, soil degradation, erosion, loss of soil organic matter (SOM) and leaching of nutrients. Therefore, sustainable concepts for increased food production are urgently needed to lower pressure on soils, in order to reduce or prevent the negative environmental impacts of intensive agriculture. A key for such strategies is the maintenance or increase of SOM level inducing positive ecosystem services such as increased productivity, nutrient and water storage, intact filter capacity, rooting, aeration and habitat for soil organism etc. (Fig. 1). In summary, SOM improves soil fertility and C storage (C Sequestration).

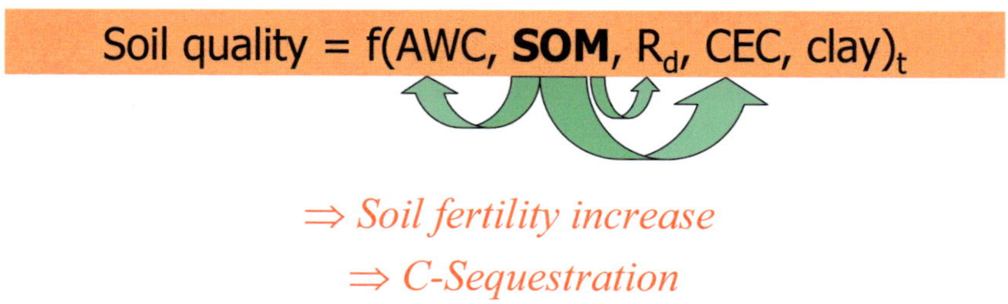

Fig. 1 Soil quality is a function (f) of available water holding capacity (AWC), soil organic matter level (SOM), root density (R_d), cation exchange capacity (CEC), clay content (clay) and time (t). The most important factor is SOM as it improves other variables such as AWC, R_d and CEC.

One efficient way to increase SOM level is compost application, produced especially from biomass wastes. During the last decades, attention was paid at the professionalization of composting due to several trends in today's society: On the one hand, growth of livestock breeding and intensification of crop production has occurred while an increasing shortage of

* Corresponding Author

resources, i.e. fossil fuels, fossil nutrients stocks and arable land, can be recognized. On the other hand, urbanization and growing population interconnected with an increased amount of waste output is responsible for environmental hazards and pollution. Therefore, composting became an efficient means of waste processing, soil amelioration and general environmental improvement.

However, up to now reported C sequestration potential due to compost management is limited in terms of C use efficiency and long-term C preservation even combined with organic farming and no till management. Therefore, new concepts for C sequestration combating against further raise of atmospheric CO_2 emissions are urgently needed. One promising option is using the "*terra preta* concept" combining biochar and composting technologies. This concept could enhance quality and material properties of compost products leading to a higher added value and to a much better C sequestration potential due to the long-term stability of biochar.

We hypothesize that composting of biochar together with other biogenous materials containing labile organic matter and nutrients can be an appropriate tool to produce a substrate with similar properties as *terra preta* such as enhanced soil fertility and C sequestration. Current available literature will be reviewed on these aspects.

2. Compost

Composting is the biological decompozition and stabilisation of organic matter derived from plants, animals or humans through the action of diverse microorganisms under aerobic conditions (Smith & Collins, 2007). The final product of this biological process is a humus-like, stable substrate, being free of pathogens and plant seeds which can be beneficially applied to land as an agent for soil amelioration or as an organic fertilizer. Although historical traditions such as those of Ancient Egyptians or Pre-Columbian Indians of Amazonia suggest that composting is an ancient method for soil amelioration, fundamental scientific studies of this biological process were published only in the past four decades. Process engineering and the knowledge about the dependence and interaction of numerous competing forces and factors within a composting matrix have been just recently established (Haug, 1993).

Multiple composting methods and systems have been developed, varying from small, home-made reactors used by individual households, over medium-sized, on-site reactors operated by farmers, to large, high-tech reactors used by professional compost producers. In spite of different process techniques, the fundamental biological, chemical and physical aspects of composting remain always the same. This concerns for example the suitability of different input materials and amendments as well as their appropriate composition, substrate degradability, moisture control, porosity, free air space, energy balance as well as decomposition and stabilization (Haug, 1993; Bidlingmaier et al., 2000).

2.1 What is compost and how is it produced?

All proper composting processes go through four stages: (1) mesophilic, (2) thermophilic, (3) cooling, finally ending with (4) compost maturation (Fig. 2). The duration of each stage depends on the initial composition of the mixture, its water content, aeration and quantity and composition of microbial populations (Neklyudov et al., 2006; Smith & Collins, 2007).

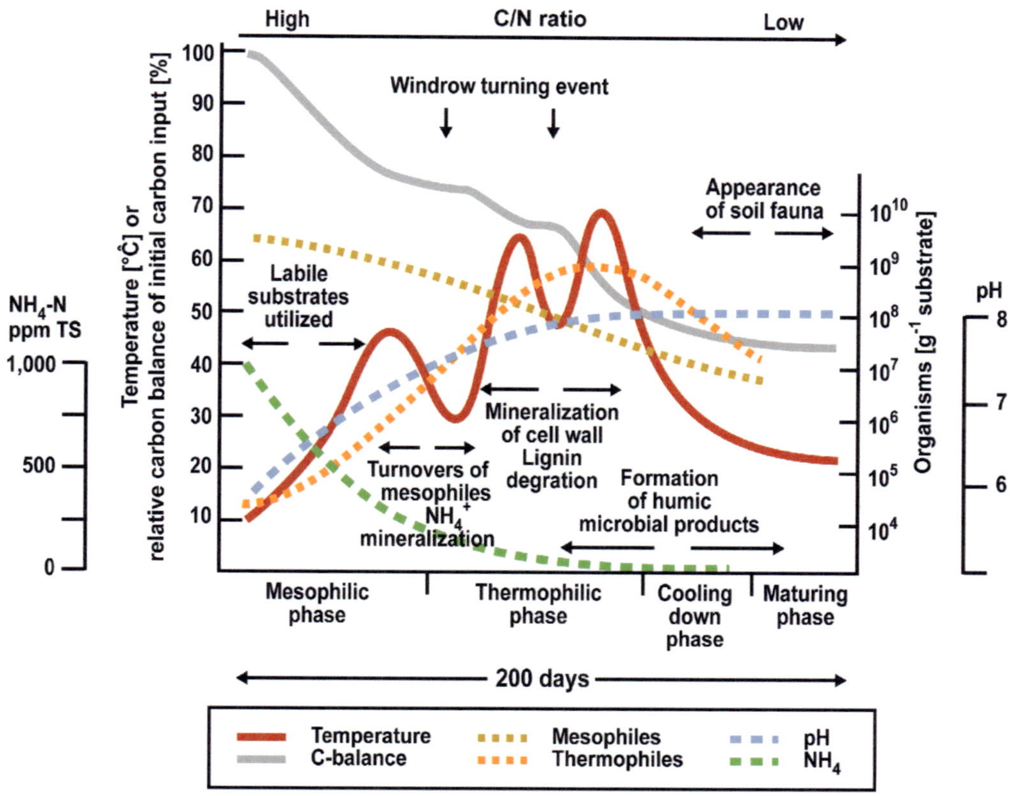

Fig. 2. Different stages during composting as function of time, appearance and succession of compost biota, temperature and further processes (based on Lechner et al., 2005 and Smith & Collins, 2007).

During the mesophilic phase, labile C-rich substrates are rapidly metabolized by a mixture of bacteria, actinomycetes and fungi preferring moderate temperature typically between 15 and 40 °C. Due to this aerobic metabolism, heat is generated. Turning the material leading to aeration temporarily decreases temperature, resulting in a rapid decomposition of further available material and thus, temperature increases again (Fig. 2). During the termophilic phase, temperature rises above 40 °C, favouring mainly actinomycetes and thermophilic bacteria such as *Bacillus*. When labile C compounds of the feed substrates decline, a gradual decrease in temperature occurs leading to the cooling phase (Fig. 2). Especially fungi have a preference for the remaining and more complex and thus degradation-resistant lignin and cellulose compounds. In addition, actinomycetes have a major importance when humic materials are formed from decomposition and condensation reactions Smith & Collins, 2007). The final maturing phase is characterized by even lower temperature below 25 °C and reduced oxygen uptake rates of aerobic microorganisms. During this stage, degradation of the more refractory organic compounds continues and soil meso and macro fauna enters. Organisms of this stage have beneficial influence on compost maturation as well as plant diseases suppression as they are able to metabolize phytotoxic compounds (Gottschall, 1984; Haug, 1993). Thus, compost quality increases especially during the last phase. Compared to the starting feed mixture, the final compost is attributed by a lower C/N ratio of 15 - 20 and a higher pH value (Smith & Collins, 2007). It can contain considerable amount of plant-available NO_{3^-} while NO_{4+} content is generally decreasing (Fig. 2). Furthermore, odour potential from compost is significantly reduced (Haug, 1993). But of utmost importance seems the fact that the organic matter has been stabilized, thus containing fairly resistant C compounds (Smith & Collins, 2007). Its application to the land influences several biological, chemical and physical soil properties in a positive and sustainable way which is outlined in Fig. 1 and which will be discussed in the following in more detail.

2.2 How does compost influence soil properties and plant growth?

Numerous publications provide evidence on the multiple benefits of compost application to soil. Effects range from soil stabilization and amelioration to phyto-sanitary impacts of mature compost. Feedstock, compost maturity and compost quality can influence intensity and degree of effects on soil physical, chemical and biological properties. Application may trigger short-term improvements such as increasing microbial activity. Long-term effects on soil properties could be achieved by preservation and increase of the stable SOM pool (Amlinger et al., 2007).

2.2.1 Can soil organic matter be maintained or increased upon compost application?

Soil organic matter is of essential importance for maintaining soil quality by improving biological, physical and chemical soil conditions (Fig. 1). It consists of a variety of simple and complex carbon compounds. While the stable SOM pool is characterized by hardly decomposable organic compounds with several beneficial, long-term effects for soil amelioration and conservation, the labile pool of SOM provides easily accessible food for soil organisms and nutrients for plant growth (Termorshuizen et al., 2005). Soil fauna, notably earthworms (*Lumbricidae*), in turn, positively influence a wide variety of different physico-chemical properties. These positive effects are especially caused by their feeding behaviour, bioturbation and the production of droppings consisting of organo-mineral complexes. The latter is of main importance for the formation of stable SOM pools in soils.

Even though the importance of SOM for the ecological functionality, efficiency and performance of soils is obvious, its worldwide reduction due to intensive agriculture gives cause for concern (Lal, 2009). According to Termorshuizen et al. (2005), this reduction has multiple reasons, but increased SOM mineralization rates due to intensive soil tillage and use of mineral fertilizers, which has often resulted in decreased application of organic soil conditioners and fertilizers, are among the most important ones.

In order to sustain agricultural productivity in the long term, SOM needs to be maintained by continuous addition of organic residues and amendments. Soil organic matter reproduction rate increases in the order green manure, leaves 15%, slurry, straw, liquid digestate 20 - 30%, fresh compost, stable manure, solid digestate 35 - 45%, mature compost > 50% (Table 1). The main factors for SOM enrichment are quantity, type and humification degree of compost and soil properties such as soil type and clay content. Mature composts increase SOM much better than fresh and immature composts due to their higher level of stable C (Bundesgütegemeinschaft Kompost e. V. [BGK] & Bundesforschungsanstalt für Landwirtschaft [FAL], 2006). There are few trials which show no significant differences in SOM level by the application of diverse C sources (straw, manure, compost). But the majority of studies of different authors have unambiguously proven a better humus reproduction for composted materials (Amlinger et al., 2007).

According to Amlinger et al. (2007), the average SOM demand of agriculturally used soils can be met by applying 7 - 10 Mg (dry matter) compost ha^{-1} a^{-1}. Therefore, for a long-term increase of SOM, however, more than 10 Mg dry matter compost ha^{-1} a^{-1} is required.

Organic fertilizer	Organic matter (dm)	Organic C (dm)	Humus reproduction	TOC increase [Mg ha^{-1}]
Green manure, leaves	90%	52%	15%	0,4
Slurry, straw, liquid manure	75%	44%	20-30%	0,5
Fresh compost, stable manure, solid digestate	50%	50%	35-45%	1,0
Mature compost	36%	50%	>50%	1,3

Table 1. Organic matter and SOM reproduction of different organic fertilizers when applied at 10 Mg ha^{-1} a^{-1} (Data from BGK & FAL, 2006).

2.2.2 How are physical soil properties influenced?

Reduction of Bulk Density: Compost application generally influences soil structure in a beneficial way by lowering soil density due to the admixture of low density OM into the mineral soil fraction. This positive effect has been detected in most cases and it is typically associated with an increase in porosity because of the interactions between organic and inorganic fractions (Amlinger et al., 2007).

Increase of aggregate stability: In general, soil structure is defined by size and spatial distributions of particles, aggregates and pores in soils. The volume of solid soil particles and the pore volume influences air balance and root penetration ability. As a general fact the more soil structure is compacted, the more unfavorable are the soil conditions for plant growth. By incorporation of compost into the soil, aggregate stability increases most effectively in clayey and sandy soils. Positive effects can be expected by well humified (promoting micro-aggregates), as well as fresh, low-molecular OM (promoting macro-aggregates). Macro-aggregates are mainly stabilized by fungal hyphen, fine roots, root hair and microorganisms with a high portion of easily degradable polysaccharides (Amlinger et al., 2007). Subsequently, the importance of stronger degraded organic compounds for the stabilization of smaller aggregates increases with time of transformation periods ranging from some few to several thousand years (Kong et al., 2005; Lützow et al., 2008; Marschner & Flessa, 2006). In this respect, the aromatic structure of inert organic compounds seems to play an important role in stabilizing micro-aggregates in the range of 2 - 20 µm as well as 20 - 250 µm by polyvalent cation bridges with clay minerals (Tisdall & Oades, 1982). Besides clay minerals and oxides, fine roots, hyphen networks as well as glue-like polysaccharides originated from root and microbial exudates significantly contribute to the formation of micro-aggregates.

Furthermore, aggregate and pore properties of soils are associated with specific "active" surface area influencing several storage and exchange processes in soil. The higher the specific surface area, the more intensive interactions can occur between soil fauna, microorganisms and root hairs under optimum conditions (e.g. sufficient humidity). As a result, a high specific surface area can create the prerequisite for an optimal soil formation (Amlinger et al., 2007).

Improvement of pore volume and hydraulic conductivity: Hydraulic conductivity is the percolation rate in soils per area and time unit depending on (i) actual soil moisture tension and (ii) number, size and form of soil pores (pore size distribution). Organic matter applied by compost improves water conductivity of soils (Carter et al., 2004) by providing a food source for soil organisms which contribute to the formation of macro-pores, in turn. Additionally, it has a direct structure-stabilizing effect for soil. A resulting increase of hydraulic conductivity is of main importance, especially for clayey soils.

Increase of field capacity, secondary pore structures and improved water retention: Field capacity (FC) is defined as the amount of water which a water-saturated soil can retain against gravity after 2 - 3 days. It is mainly influenced by pore volume and pore size distribution because only pores below a pore diameter of 50 µm (corresponding to pF 1.8) can retain water against gravity due to higher capillary force. However, adhesive force of pores with a diameter below 0.2 µm is so high that this water is not available for plants (permanent wilting point or PWP corresponding to pF 4.2). Field capacity and available water holding capacity (AWC, pF 1.8 - 4.2) are generally influenced by the particle size, structure and content of OM. Several studies confirm a significant, positive impact of OM amendment to soils on FC (Evanylo et al., 2008; Tejada et al., 2006; Carter et al., 2004). Amongst others, this effect results from the improved formation of secondary pore structures which can be mainly ascribed to root and animal tubes. This is important for soils with low portions of primary meso-pores. In this respect, compost increases the portion of meso- and macro-pores because of an improved aggregation and stabilization of soil significantly initiated by various soil organisms (Liu et al., 2007). In addition, organic matter (OM) is able to take up 3 to 20 times more water compared to its own weight. Considering these effects, an increase of total organic carbon (TOC) content from 0.5% to 3% resulted in a duplication of AWC (Hudson, 1994).

Improved air balance: The portion of air in soils results from the difference between total pore volume and the pores filled with water (Amlinger et al., 2007). Air permeability and air exchange in soils predominantly depends on pores > 50 µm. While sandy soils are characterized by a high portion of primary macro pores resulting in a proper aeration, clayey or compacted soils have few macro pores which may cause lack of oxygen availability. For these latter soils, OM applied by compost has a significant ameliorating effect by improving porous soil structure and its stabilization, stimulating the formation of secondary macro pores especially by roots and animals tubes.

Reduction of soil erosion and run-off: Reduced erosion is mainly related to the improved soil structure by the addition of compost which, in turn, is pointed out by better infiltration rate, pore volume and enhanced stability through aggregation (Diacono & Montemurro, 2010). According to Amlinger et al. (2007), experimental trials showed a clear correlation between increases of SOM, reductions of soil density, soil loss and water run-off. The effect of compost on soil erosion has been quantified in detail by Strauss (2003). Five years long compost application resulted in 67% reduced soil erosion, 60% reduced run-off, 8% lower bulk density and 21% higher OM content compared to control plots. Similar results were observed by Hartmann (2003) in a wind tunnel experiment by testing the resilience of compost application against wind erosion for two different soil types: By the incorporation of compost, loss of soil particles from the topsoil was reduced to a maximum of 61 % for a podzol and 71 % for a luvisol.

Improved heat balance of soils: Soil temperature influences the reaction rate of chemical, metabolic and biological growth processes of organisms. While temperature fluctuations mainly depend on climate, radiation absorption can be influenced by color. Composts are dark-colored resulting in higher light absorption and thus lower albedo (reflection rate of light from a light source). Thus, higher light absorption will warm up soils supplied with compost faster than light-colored soils (Stöppler-Zimmer et al., 1993). This will promote germination of seeds, especially during spring. However, as temperature increases in summer, uncovered dark soils can heat up extremely. As a result, soil can dry out due to a higher evaporation which, in turn, affects plant growth and soil biology negatively. In such case, compost mulching systems offer a good solution because they obtain reduced fluctuations of soil temperature which results from the shading effect of mulch loosely covering the soil surface.

2.2.3 How are chemical soil properties influenced?

Enhancement of nutrient level: Compost contains significant amounts of valuable plant nutrients including N, P, K, Ca, Mg and S as well as a variety of essential trace elements (Seiberth & Kick, 1969; Bischoff, 1988; Lenzen, 1989; Haug, 1993; Smith & Collins, 2007). Thus, compost can be defined as an organic multi nutrient fertilizer (Hartmann, 2003; Amlinger et al., 2007). Its nutrient content as well as other important chemical properties like C/N ratio, pH and electrical conductivity (EC) depend on the used organic feedstocks and compost processing conditions (Table 2). By an appropriate mixture of these organic input materials humus and nutrient-rich compost substrates can be produced (Table 3) serving as a substitute for commercial mineral fertilizers in agriculture. However, their diverse beneficial properties for amelioration outreach their nutrient content.

	TOC	TN	C/N ratio	pH_{H2O}	Total P	Total K	Total Ca	Total Mg	Reference
	--- $g\ kg^{-1}$ DM ---				------------- $g\ kg^{-1}$ DM -------------				
Household waste	368	21.7	17	4.9					Eklind et al. (1997).
Manure	330	22	15	9.4	3.9	23.2	9.1	4.8	Kimetu et al. 2008
Wood chips	394	14.3	28	7.4	3.5				Larney et al. 2008
Sawdust	490	1.1	446	5.2	0.1	0.4	1.5	0.1	Kimetu et al. 2008
Straw	358	1.0	138	8.3	4.0				Beck-Friis et al. (2001); Larney et al. 2008
Canola	457	1.9	24	6.3	1.1				Yuan et al. 2011
Rice	412	8.7	47	6.8	1.1				Yuan et al. 2011
Soybean	440	23.8	18	6.3	0.9				Yuan et al. 2011
Pea	436	35.0	12	6.3	4.6				Yuan et al. 2011

DM = dry matter
TOC = total organic carbon
TN = total nitrogen
EC = electrical conductivity

Table 2. Chemical properties of organic feedstock materials.

However, total nutrient content of compost is not plant-available to the full extent at once. This can be ascribed to the existence and different intensity of various binding forms within the organic matrix which result in a partial immobilization of nutrients (Becker et al., 1995). On the one hand, this condition makes it more difficult to calculate the fertilization effect and to estimate the nutrient balance in advance (Becker et al., 1995). On the other hand, the fertilization effect will last longer due to a slow and gradual release of plant nutrients (Smith & Collins, 2007). Therefore, with compost there is a much better protection from leaching compared to soluble mineral fertilizers. Especially the N fertilization effect of compost is limited due to

Table 3. Chemical properties of compost products.

	Water content	OM	TOC	TN	C/N ratio	pH_{H2O}	EC (1:5)	NH_4^+–N	NO_3^-–N	Total P	Total K	Total Ca	Total Fe	Refference
	%	% DM	--- g kg^{-1} DM ---				dS m^{-1}	--- mg kg^{-1} DM ---		--- % DM ---			g kg^{-1}	
Municipal solid waste	42.1	48.6	252	17	16.4	5.9	10.9	2100						Alluvione et al, 2010; Annabi et al, 2007; Vaz-Moreira et al. 2007
Bio-waste		36.2	181	19	9.5	8.5								Annabi et al, 2007
Domestic waste	36.2	38.3	192	15	12.8	6.8	0.5	<LD						Vaz-Moreira et al, 2007
Vermicompost	9.5	31.4	157	13	12.1	6.5	8	70						Vaz-Moreira et al, 2007
Green waste	37.9		291	0.8	38.3			41	135					Dalal et al, 2009
Green waste + Sludge		48.5	242	24	10.1	6.6								Annabi et al, 2007
Sewage Sludge	59.6	52.6	319	28	18.1	4.8	6.5	1030	51.5	1.3	1.9	0.1	12.8	Ahmad et al, 2008; Bar-Tal et al, 2004; Beraud et al, 2005; Vaz-Moreira et al. 2007
Manure														
Farmyard manure		36.2	210	11	19.1	8		40	120	0.5	0.5			Badr EL-Din et al. 2000
Feedlot manure	28.2		313	2.6	12			4.2	0					Dalal et al, 2009
Poultry	24.7	46.8	278.5	33.9	8.4	8.3	9.2	2900		0.6	4	6	0.1	Ahmad et al, 2008; Vaz-Moreira et al, 2007
Cattle		25.2	225	18	13.4	7.6	8	32.5	946.5	0.8	2.9	0.6	3.3	Ahmad et al, 2008; Bar-Tal et al, 2004; Beraud et al, 2005

DM = dry matter
OM = organic matter
TOC = total organic carbon
TN = total nitrogen
EC = electrical conductivity
LD = limit of detection (=0,04 mg kg-1)

low mineralization rates and microbial immobilization (Kehres, 1992, Vogtmann et al., 1991). In the first year after compost application, only 10 - 20% of the total N content will be mineralized according to Becker et al. (1995). Bidlingmaier et al. (2000) reported N mineralization rate of 10% in the first year and 40% in total in the long term. In contrast to the low N availability, a higher fertilization effect for P (50%), K (100%) and Mg could be observed (Bidlingmaier et al., 2000). With respect to micro nutrients, an increased plant uptake of Cu, Mn and Zn was reported (Amlinger et al., 2007).

Increase of cation exchange capacity (CEC): The CEC is one of the most important indicators for evaluating soil fertility (Fig. 1), more specifically for nutrient retention and thus it prevents cations from leaching into the groundwater. Kögel-Knabner et al. (1996), Kahle & Belau (1998) and Ouedraogo et al. (2001) proved that compost amendment resulted in an increase of CEC due to input of stabilized OM being rich in functional groups into soil. According to Amlinger et al. (2007), SOM contributes about 20 - 70% to the CEC of many soils. In absolute terms, CEC of OM varies from 300 to 1,400 $cmol_c\ kg^{-1}$ being much higher than CEC of any inorganic material.

Increase of pH value, liming effect and improved buffering capacity: Soil pH is is an indicator for soil acidity or soil alkalinity and is defined as the negative logarithm of hydrogen ions activity in a soil suspension. It is important for crop cultivation because many plants and soil organisms have a preference for slight alkaline or acidic conditions and thus it influences their vitality. In addition, pH affects availability of nutrients in the soil. Compost application has a liming effect due to its richness in alkaline cations such as Ca, Mg and K which were liberated from OM due to mineralization. Consequently, regularly applied compost material maintains or enhances soil pH (Kögel-Knabner et al., 1996; Diez & Krauss, 1997; Kahle & Belau, 1998; Stamatiadis et al., 1999; Ouedraogo et al., 2001). Only in some few cases a pH decrease was observed after compost application (Zinati et al., 2001).

Reduction and immobilization of pesticides and persistent organic pollutants (POPs): A contamination of soils or composts with pesticides and POPs can occur in consequence of environmental pollution, conventional farming practice by using chemicals and pesticides and by incorporation of contaminated materials into compost or soil. Therefore, unpolluted feedstocks should be generally preferred for composting in order to avoid critical concentrations of pollutants. However, pesticides and POPs can be degraded or immobilized during compost processing or by the properties of the final compost product. Based on temperature and oxidative microbial and biochemical processes, composting contributes to an effective reduction of organic pollutants. For instance, polychlorinated biphenyls (PCB) were degraded up to 45% during composting (Amlinger et al., 2007). Linear alkylbenzene sulphonates (LAS), Nonylphenols (NPE) and Di (2-ethylhexyl) phthalate (DEHP) which are mainly found in sewage sludge, are degraded almost completely under oxidative conditions (Amlinger et al., 2007). Furthermore, the degradation rate of halogenated organic compounds and pesticides is much higher than in soils, especially during the thermophilic stage (Amlinger et al., 2007). Mineralization rate of pollutants is reported to be more effective in compost soil mixtures if mature compost is applied. This concerns especially the degradation of polycyclic aromatic hydrocarbons (PAH) and other hydrocarbons (Amlinger et al., 2007). Due to the high level of humified OM, particularly mature composts contribute to sorption and immobilization of POPs resulting in a lower availability of POPs and reduced toxicity.

Immobilization of heavy metals: Similar to pesticides and POPs, there are several sources for heavy metal input. To a limited extent, heavy metals or trace elements serve as plant nutrients while their accumulation can cause toxicity. In this respect, OM applied by compost is able to adsorb heavy metals and reduce their solubility resulting in immobilization. Apart from some non-crystalline minerals with very high surface areas SOM has probably the greatest capacity to bond most heavy metals (Amlinger et al., 2007). The sorption strength of heavy metals to SOM generally decrease in the following order: Cr(III) > Pb(II) > Cu(II) > Ag (I) > Cd (II) = Co(II) = Li(II). On the other hand, significant correlations between the solubility of Cd, Cu, Zn, Pb and Ni and SOM content have been reported by Holmgren et al. (1993) for a range of soils from the USA. Organic matter applied by compost even effectively prevents mobilization of heavy metals for a long time after the cessation of compost addition (Leita et al., 2003). In order to guarantee the lowest possible pollutant input over time, raw materials for composting have to be separated and preferably unpolluted feedstocks should be used. However, the prevention of environmental pollution and thus the contamination with heavy metals is basically a general matter for the society and for politics. As long as there is an emission of pollutants by industry and society immission will occur including the pollution of valuable organic feedstocks.

2.2.4 How are biological soil properties influenced?

One of the most important effects of compost use is the promotion of soil biology. In this respect, the following three aspects seem essential: (i) Food supply for soil heterotrophic organisms by adding degradable carbon compounds with OM (Blume, 1989); (ii) optimization of habitat and niche properties in soil, e.g. water and air balances, increase of specific surfaces, retreat areas etc.; (iii) introducing compost biota into soil as an inoculant (Amlinger et al., 2007; Sahin, 1989; Werner et al., 1988). Compost has a stimulation effect on both the microbial community in the compost substrate as well as the soil-born microbiota of soils (Table 4). Two fractions of OM are responsible for the level of microbial activity in general: (i) Easily degradable organic compounds (labile OM pool) may increase microbial activity and biomass temporarily while (ii) a persisting increase of microbial biomass depends on a constant enhancement of stable OM which is particularly promoted by mature compost addition.

Material	Bacteria [10^6 g^{-1} dm]	Fungi [10^3 g^{-1} dm]
Pesticide containing soil	19	6
Reclaimed soil after surface mining	19-70	8-97
Fertile soil	6-46	9-46
Mature green waste compost	417	155

Table 4. Soil bacterial and fungal biomass in soils and compost (United States Environmental Protection Agency - Solid Waste and Emergency Response, 1998).

Microorganisms perform several ecological and environmental functions. With regard to compost, the following microbial effects seem of main importance:

- Degradation and humification: A gradual breakdown of organic compounds is performed by a succession of different soil organisms over time. While at the beginning easily degradable organic substances are decomposed, further decomposition and transformation of the remaining by-products occur, finally resulting in a stable humus-like compost product which is subjected to only slow decomposition rates.

- Mineralization, biological immobilization and nutrient cycling: On the one hand, microorganisms convert complex organic substances to low-molecular, inorganic substance. By this mineralization process, nutrients are released for plant growth so that the plant nutrients can cycle within the ecosystem. On the other hand, soil organisms immobilize nutrients into their own biomass. By this way, e. g. N is protected from leaching.
- Aggregation: Microorganisms contribute to the formation and stabilization of aggregates by the synthesis of biofilms and exudates as well as by their living or dead biomass.
- Degradation or reduction of pesticides, POPs and phytotoxic compounds: By microbial metabolism, several chemical compounds which are harmful for plants, can be decomposed, transformed or immobilized.
- Suppression of pathogens and diseases: The diversity of microorganisms in mature composts exhibit suppressive effects on several pathogens and diseases which could harm plant life or human health (Amlinger et al., 2007).

2.2.5 How are plant growth, plant health and crop quality influenced?

In general, compost creates a favourable environment for plant and root growth especially by

1. promoting a porous soil structure for optimized root penetration;
2. decreasing soil erodibility due to the formation of stable aggregates. Consequently, plant roots are less exposed to direct damage caused by eroded topsoil and water can better infiltrate into soil. Furthermore, air exchange is less interfered by the compaction of subsurface soil or by the formation of a soil crust, which tends to "seal" the surface (Buchmann, 1972; Richter, 1979; Krieter, 1980; Fox, 1986; Löbbert & Reloe, 1991);
3. intensifying essential interactions between root hairs, soil fauna and microorganisms due to an enhancement of specific surface area (Amlinger et al., 2007);
4. improving percolation. On the one hand, this prevents waterlogging which can result in a decay of plant roots due to anaerobic soil conditions. On the other hand, loss of nutrients is reduced by decreased run-off;
5. enhancing water storage capacity and improving water retention which helps plants better overcome critical climate conditions like droughts (Hartmann, 2003);
6. providing valuable macro- and micro-nutrients in the long term (Gottschall, 1984) due to slow mineralization rates, better nutrient adsorption as well as enhanced storage capacity which prevents from leaching;
7. improving buffering capacity which helps to maintain uniform reactions and conditions for better plant growth;
8. promoting the degradation, reduction or immobilization of harmful substances like pesticides, POPs, heavy metals and phyto-toxic compounds which can interfere plant life and health;
9. providing microbial symbionts and beneficial soil organisms a habitat which, in turn, has a positive influence on vitality and growth of plants;
10. protecting plants from pathogens and diseases due to antiphytopathogenic potential of compost (Hoitink, 1980; Nelson & Hoitink, 1983; Hoitink & Fahy, 1986; Blume, 1989; Hadar et al., 1992; Bidlingmaier et al., 2000);

Due to its multiple positive effects on the physical, chemical and biological soil properties, compost contributes to the stabilization and increase of crop productivity and crop quality (Amlinger et al., 2007). Long-term field trials proved that compost has an equalising effect of annual/seasonal fluctuations regarding water, air and heat balance of soils, the availability of plant nutrients and thus the final crop yields (Stöppler et al., 1993; Amlinger et al., 2007). For that reason, a higher yield safety can be expected compared to pure mineral fertilization. Better crop results were often obtained if during the first years higher amounts of compost were applied every 2nd to 3rd year than by applying compost in lower quantities of < 10 Mg (DM) ha^{-1} every year (Amlinger et al., 2007). However, crop yields after pure compost application were mostly lower when compared to mineral fertilization (Amlinger et al., 2007), at least during the first years. This can be explained by the slow release of nutrients (especially nitrogen) during mineralization of compost.

Compost use does not only improve the growth and productivity of crops in terms of quantity but it could be also proved that quality of agricultural products is influenced in a positive way (Söchtig, 1964; Flaig, 1968; Harms, 1983). By examination of several crops in situ, an increase of beneficial and healthy ingredients and a decrease of harmful substances in the final crop product after compost use compared to a treatment with mineral fertilizer application were observed (Vogtmann et al., 1991). In addition, Fricke et al. (1990) reported significantly higher dry matter content of beet root as well as a lower nitrate level after compost amendment compared to mineral fertilized sets. In a second trial with potatoes, the same authors detected a higher content of starch, vitamin C and dry matter for the compost-treated plants compared to the mineral fertilization variant. Furthermore, the portion of marketable potato tubers with respect to the total yield was enhanced in the compost treatment.

In spite of the potential and observed beneficial effects of compost application to agricultural soils, this technique is not widespread across Europe and especially in Germany, low quality composts are produced due to inefficient waste management regulations. In addition, long-term C sequestration potential of compost remains insufficient with respect to mitigation of global atmospheric CO_2 increase. Furthermore, as presented by data in Table 5, CO_2 emission during the rotting process of compost production generally

	Fresh compost from bio-waste			Mature compost from bio-waste			Mature compost from yard + parc waste		
	Processing	CO2 emission	Screening	Processing	CO2 emission	Screening	Processing	CO2 emission	Screening
Feedstocks	100			100			100		
loss during rotting process		35			55			55	
compost, not sieved	65			45			45		
compost, sieved	43			25			16		
Screening remains			22			20			29

Table 5. Relative carbon balance in percent of initial carbon input based on data from Reinhold (2009) concerning composting facilities in Germany.

cause high total carbon losses of 35 - 55 % compared to the initial carbon input by organic feedstocks. This fact indicates that current composting practice may be optimized with respect to a more efficient carbon conservation. Therefore, additional concepts such as *terra preta* / biochar are required which will be discussed in the following.

3. Biochar

3.1 How can biochar improve soils?

In central Amazonia, up to 350 ha wide patches of a pre-Columbian black earth-like anthropogenic soil exist, very well known as *terra preta (de Indio)* characterized by a sustainable enhanced fertility due to high levels of SOM and nutrients such as N, P and Ca (Glaser et al., 2001; Glaser, 2007; Glaser & Birk, 2011). However, the key for *terra preta* formation is the tremendous input of charred organic materials, known as biochar comprising up to 35% of SOM and on average 50 Mg ha^{-1} (Glaser et al., 2001). Biochar acts as a stable C compound being degraded only slowly with a mean residence time in the millennial time scale. Biochar has a high specific surface area (400 - 800 m^2 g^{-1}), it provides a habitat for soil microorganisms which can degrade more labile SOM. In addition, higher microbial activity accelerates soil stabilization as outlined in the previous section. Furthermore, higher mineralization of labile SOM and biochar itself provided important nutrients for plant growth. The general recipe of *terra preta* generation and the principal function of biochar are shown in Fig. 3.

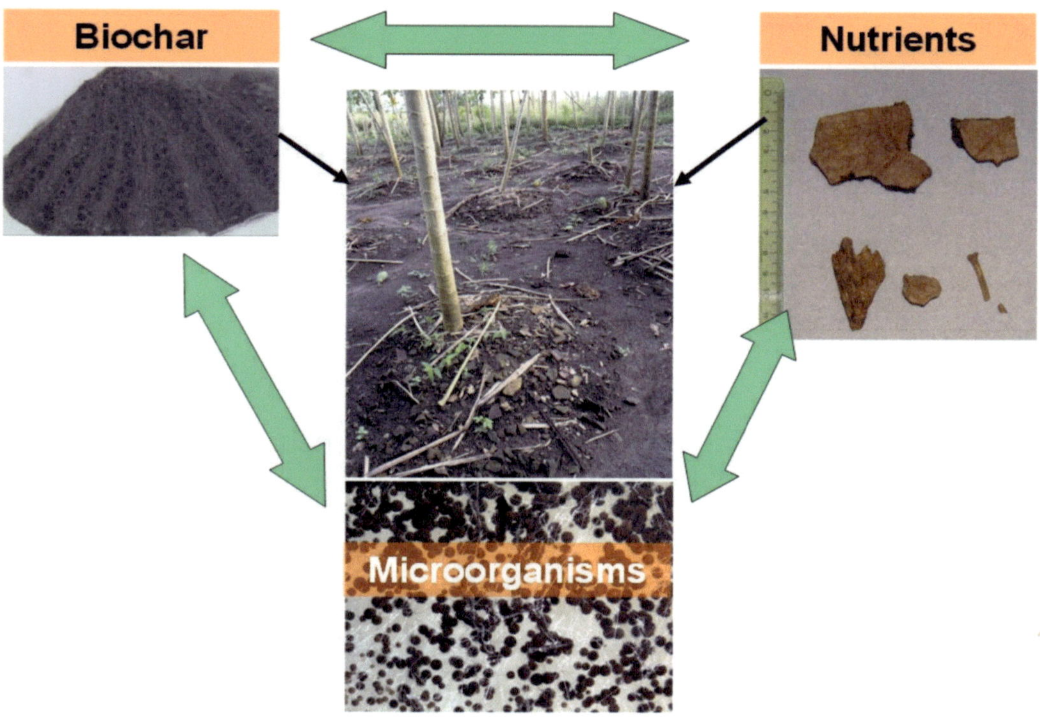

Fig. 3. Principles of terra preta formation and soil biochar interaction.

3.2 What is biochar exactly?

Biochar is produced by thermal treatment at oxygen deficiency e. g. by pyrolysis or gasification, resulting in three products: char, gas and tarry oils. The relative amounts and characteristics of each are controlled by the process conditions such as temperature, residence time, pressure, and feedstock type. Biochar production can be chemically described by water elimination followed by increasing aromatic condensation, which can be expressed as decreasing atomic ratios of O/C and H/C along the combustion continuum (Fig. 4). However, biochar is no clearly defined chemical compound. Instead, it is a class of compounds along the combustion continuum and we need to define thresholds for materials which are claimed to be biochar (Fig. 4). Recently, on the basis of about 100 biochar samples differing in feedstock and production process, the following elemental ratio thresholds were suggested for biochar, $O/C < 0.4$ and $H/C < 0.6$ (Schimmelpfennig & Glaser, 2012).

As a consequence, the content of condensed aromatic moieties, known as black carbon, increases being responsible for its stability in the environment. The second important ecological property of biochar is presence of functional groups on the edges of the polyaromatic backbone (Fig. 4) which are formed by partial oxidation (Glaser, 2007). Therefore, biochar is an option for long-term C sequestration while maintaining or increasing soil fertility which was successfully proven by the *terra preta* phenomenon for at least 2,000 years (Glaser, 2007). Due to this fact, *terra preta* could be a model for sustainable resource management in the future not only in the humid tropics but also in temperate and arid regions around the world providing a solution for land degradation due to intensive land use and growing world population. In the following, we will review reported biochar effects on ecosystem services.

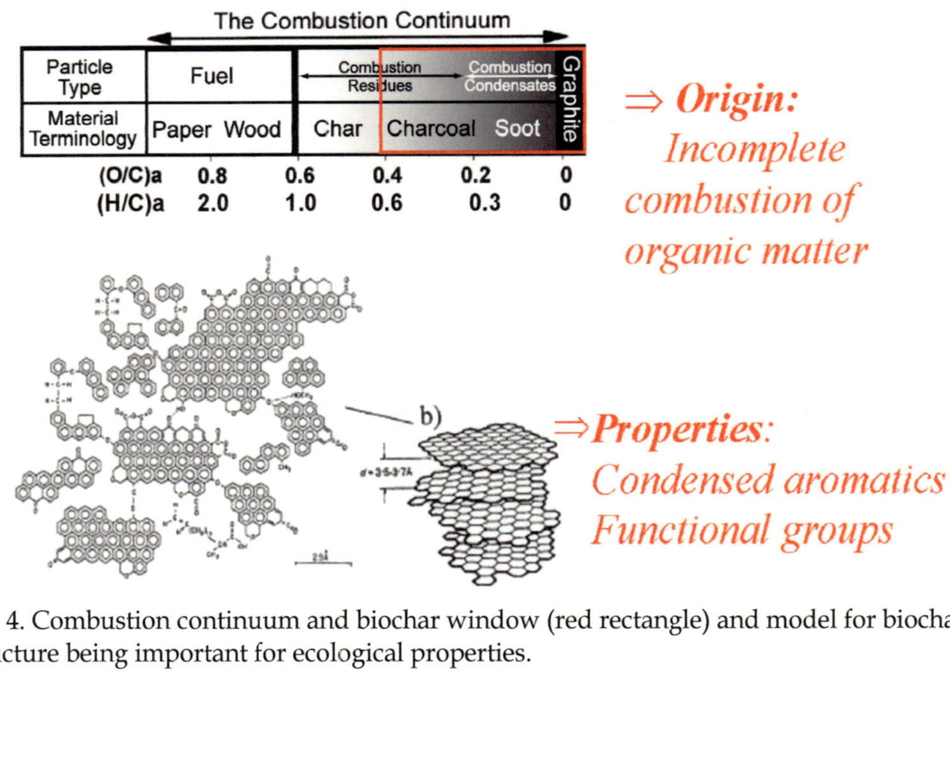

Fig. 4. Combustion continuum and biochar window (red rectangle) and model for biochar structure being important for ecological properties.

3.3 How long will biochar survive in soil?

Due to its recalcitrance against microbial degradation, biochar is very stable in soil compared to other OM additions, making its application to soils a suitable approach for the build-up of SOM and thus, for C sequestration. The prevailing scientific understanding of biochar degradation in soil is that some portions of it are quite readily decomposable (labile), while the core structure of the material is highly resistant to degradation (Fig. 4). Biochar in *terra preta* has been dated to 1,000 to 1,500 years (Glaser et al., 2000) and naturally occurring biochar in Australian soils to 1,300 - 2,600 years (Lehmann et al., 2008). As SOM decomposition rates in temperate regions are slower, mean residence time for biochar can be assumed to be higher in European soils. Controlled biochar decomposition experiments revealed a mean residence time in soils between 1,300 to 4,000 years (Cheng et al., 2008; Liang et al., 2008; Kuzyakov et al., 2009). Management practices such as tillage and addition of labile C (e.g. slurry) to soil significantly increased biochar mineralization by a factor of 0.5 to 2, however, only in the short-term (Kuzyakov et al., 2009) so that biochar application can be combined with such agricultural technologies without the disadvantage of additional SOM and biochar degradation.

In a range of other biochar incubation experiments, the interactive effects of biochar addition to soil on CO_2 evolution (priming) were evaluated by comparing the additive CO_2 release expected from separate incubations of soil and biochar with corresponding biochar and soil mixtures. Positive (C mineralization stimulation) or negative (C mineralization suppression) priming effects and magnitude varied with soil and biochar type. In general, C mineralization was higher than expected (positive priming) for soils combined with biochars produced at low temperatures (250 - 400 °C) and from grasses, particularly during the early incubation stage (first 90 d) and in soils of lower organic C content (Zimmerman et al., 2011). In contrast, C mineralization was generally less than expected (negative priming) for soils combined with biochars produced at high temperatures (525 - 650 °C) and from hard woods, particularly during the later incubation stage (250 - 500 d). Obtained data strongly suggests that biochar soil interaction will enhance C sequestration via SOM sorption and organo-mineral interaction in the long term.

3.4 How much biochar can be stored in soils?

C sequestration with biochar addition to soils could be quite significant since the technology could potentially be applied in many areas including croplands, grasslands and also a fraction of forestlands. The maximum capacity of carbon sequestration through biochar soil amendment in croplands alone was estimated to be about 428 Gt C for the world (Table 6). This capacity is estimated according to (i) the maximal biochar amount that could be cumulatively placed into soil while still beneficial to soil properties and plant growth; and (ii) the arable land area that the technology could potentially be applied through biochar agricultural practice. If using also grassland soils and 30% of forest soils, a worldwide biochar sequestration potential of 1,126 Gt C would be possible (Table 6).

3.5 Can we solve our climate problem with biochar alone?

Photosynthesis captures more CO_2 from the atmosphere than any other process on Earth. Each year, terrestrial plants photosynthetically fix about 440 Gt CO_2 being equivalent to 120

Applicable world lands	Land area /million hectares	Estimated capacity (GtC) of biochar carbon storage in soil
Croplands	1411	428
Temperate grasslands	1250	380
30% Forest lands	1181	358
Total	3842	1166

Table 6. Potential C sequestration with biochar in soils of the world. Estimated storage capacity is based on a maximum of 10 weight% biochar addition to the upper 30 cm and a soil bulk density of 1.3 Mg m^{-3} and 70% stable carbon in biochar (Lee et al., 2010).

Gt C per year from the atmosphere into biomass (Smith & Collins, 2007). This corresponds to about one-seventh of the CO_2 stock in the atmosphere (820 Gt C). However, biomass is not a stable form of carbon material with nearly all returning to the atmosphere in a relatively short time as CO_2 because of respiration and biomass decomposition. As a result, using biomass for carbon sequestration is no good option. Any technology that could significantly prolong the lifetime of biomass materials would be helpful to global carbon sequestration. A conversion of only about 7% of the annual terrestrial gross photosynthetic products into a stable biomass carbon material such as biochar would be sufficient to offset the entire amount (nearly 9.7 Gt C a^{-1}) of CO_2 emitted into the atmosphere annually from the use of fossil fuels (www.iwr.de). More realistic estimates are that annual net CO_2, CH_4 and N_2O emissions could be reduced by a maximum of 1.8 Pg C a^{-1} without endangering world food security and soil fertility (Woolf et al., 2010) corresponding to 16% of current anthropogenic CO_2 emissions. Therefore, biochar can significantly contribute to climate change mitigation but additional technologies are required to quantitatively offset fossil fuel-derived CO_2 emissions. The substitution of fossil fuels by developing and extending renewable energies is another essential key factor for greenhouse gas emission reduction or avoidance while still meeting the basic requirements of electrical or thermic energy consumption demands of society. An impressive example for such a concept is the integrative combination of innovative technologies: a PYREG pyrolysis reactor unit (Gerber 2010), for instance, locally produces biochars from organic wastes in an environmentally friendly way, while also generating heat and electricity from renewable, carbon neutral resources.

Thus, a decentralized application of this technology represents a promising and sustainable strategy for the future.

3.6 Can biochar increase soil fertility and thus crop performance?

Biochar application to soil influences various soil physico-chemical properties. Due to the high specific surface area of biochar and because of direct nutrient additions via ash or organic fertilizer amendments, nutrient retention and nutrient availability were reported being enhanced after biochar application (Glaser et al., 2002; Pietikäinen et al., 2000). Higher nutrient retention ability, in turn, improves fertilizer use efficiency and reduces leaching (Steiner et al., 2008; Roberts et al., 2010). Most benefits for soil fertility were obtained in highly weathered tropical soils but also higher crop yields of about 30% were obtained upon biochar addition in temperate soils (Verheijen, 2009). Furthermore, enhanced water-holding capacity can also cause a higher nutrient retention because of a reduced percolation of water and the herein dissolved nutrients (Glaser et al., 2002).

However, since biochar has only low nutrients contents in general, plant nutrients must be supplied externally (Woods & Mann, 2000; Glaser & Birk, 2011). With respect to potential nutrient sources, only C and N can be produced in situ via photosynthetic organisms and biological N fixation, respectively. All other elements, such as P, K, Ca and Mg must be added for nutrient accumulation (Glaser, 2007) which can be best achieved by adding organic fertilizers such as manure or compost (Schulz & Glaser, 2011).

3.7 The role of soil organisms

According to Ogawa (1994), biochar is generally characterized by a proliferation effect for several symbiosis microorganisms due to its porous structure providing an appropriate habitat for soil microbes. Steiner et al. (2004) observed a significant increase of microbial activity and growth rates by applying biochar to a Ferralsol. Furthermore, an increase of soil microbial biomass and a changed composition of soil microbial community were also observed after biochar amendments (Birk et al., 2009).

While microbial reproduction rates after glucose addition in soils amended with biochar increased, soil respiration rates were not higher (Steiner et al., 2004). This difference between low soil respiration and high microbial population growth potential is one of the characteristics of *terra preta*. These results indicate that a low biodegradable SOM together with a sufficient soil nutrient content are able to support microbial population growth. According to Birk et al. (2009), these effects can be ascribed to different habitat properties in the porous structure of biochar. The following factors might be the reason in decreasing order of currently available evidence supporting them:

- high surface area and porous structure of biochar suitable for several kinds of microbes as habitat and retreats;
- enhanced ability to retain water and nutrients resulting in a stimulation of microbes;
- formation of 'active' surfaces covered by water film, dissolved nutrients and substances providing an optimal habitat for microorganisms; these specific surfaces serve as interaction matrix for storage and exchange processes of water and substances between soil fauna, microorganisms and root hairs (Amlinger et al., 2007);
- weak alkalinity (Ogawa, 1994);
- preserving character against decay probably resulting in the (partial) inhibition of certain 'destructive' and pathogenous organisms while simultaneously supporting beneficial microbes.

Based on these possible stimulating factors, biochar promotes the propagation of useful microorganisms such as free-living nitrogen fixing bacteria (Tryon, 1948; Ogawa, 1994; Nishio, 1996; Rondon et al., 2007). Further reports from Japanese scientists prove increased yields by the stimulation of indigenous arbuscular mycorrhizal fungi (AMF) via the application of biochar (Ogawa, 1994; Nishio, 1996; Saito & Marumoto, 2002): e. g. improved yields for soybeans because of enhanced nodule formation by means of biochar addition (Ogawa et al., 1983). Results from Nishio (1996) obtained with alfalfa (*Medicago sativa*) in pot experiments indicate that biochar was ineffective in stimulating alfalfa growth when added to sterilized soil. However, alfalfa shoot weight was increased by a factor of 1.7 and nodule weight by 2.3 times in a treatment receiving biochar, fertilizer and rhizobia compared to a set only treated with fertilizer and rhizobia. According to Nishio (1996), this clearly indicates that the stimulatory effect of adding biochar may appear only when a certain level of indigenous AMF is present. Biochar amendment generally seem to stimulate soil fungi which seems logic as biochar is a complex matrix being degradable only by soil fauna and soil fungi (Birk et al., 2009).

4. Combined compost and biochar

4.1 Effect of process (mixing, composting, fermentation)

Terra preta was most likely formed by mixing of charring residues (biochar) with biogenic wastes from human settlements (excrements and food wastes including bones and ashes) which were microbially converted to a biochar-compost-like substrate (Glaser et al., 2001; Glaser, 2007; Glaser and Birk, 2011). Thus, co-composting of biochar and fresh organic material is likely to have a number of benefits compared to the mere mixing of biochar or compost with soil. Examples are enhanced nutrient use efficiency, biological activation of biochar and better material flow management and a higher and long-term C sequestration potential compared to individual compost and biochar applications (negative priming effect).

Compared to compost and biochar mixing, an increased decomposition of biochar can be expected during composting although biochar is much more stable than other organic materials. As observed by Kuzyakov et al. (2009), biochar decomposition rates increase as long as easily degradable C-rich substrate is available. Additionally, Nguyen et al. (2010) reported that higher temperature increased biochar oxidation and thus decomposition. However, these effects are much lower for biochar than for compost feedstock. On the other hand, surface oxidation will enhance the capacity of biochar to chemisorb nutrients, minerals and dissolved OM. The overall reactivity of biochar surfaces therefore probably increases with composting (Thies & Rillig, 2009).

From the compost point of view, there is evidence that biochar as a bulking agent improves oxygen availability and hence stimulates microbial growth and respiration rates (Steiner et al., 2011). Pyrolysis condensates adsorbed to biochar initially provoked increased respiration rates in soils which most likely occur also during composting (Smith et al., 2010). Biochar in compost provides habitats for microbes, thereby enhancing microbial activity. Steiner et al. (2011) reported increased moisture absorption of biochar-amended composts with beneficial effects on the composting process.

It was often stated in non-scientific literature, that *terra preta* was formed by anaerobic fermentation of biochar with organic wastes using "effective microorganisms ®" (EM ®) which consist mainly of a mix of lactic acid and photosynthetic bacteria, yeasts, actinomycetes as well as other genera and species of beneficial microorganisms (Higa & Wididana, 1991). However, there is no scientific proof for this and from a practical point of view it is most unlikely that pre-Columbian Indians manually moved tremendous amounts of soil and organic wastes for fermentation in closed containers. For the average dimension of *terra preta* being 20 ha wide and one meter deep, 200,000 m^3 or 260,000 tons of soil would have being moved by hand twice (forth and back) for terra preta generation which is most unlikely.

Nevertheless, fermentation theoretically provides microorganisms to soil which could be beneficial for soil health and ecosystem services. In a composting / fermentation experiment with and without biochar, the overall C loss during fermentation was about 30% lower when compared to composting (Fig. 5). However, when composting the fermented material, overall OC loss was even higher compared to the composted only material (Fig. 5). This indicates that fermented OM is only stable as long as it was kept anaerobic. As soon as piles were turned (after fermentation) and oxygen became available, the intermediate fermentation products were mineralized to an even higher extent than the non-fermented counterparts (Fig. 5). Biochar addition appeared to amplify fermentation-induced stabilization, since compost piles with 50 (DEM50) and 100 (DEM100) kg biochar per ton of organic feedstock material showed reduced OC loss compared to the fermentation control without biochar (DEM0, Fig. 5).

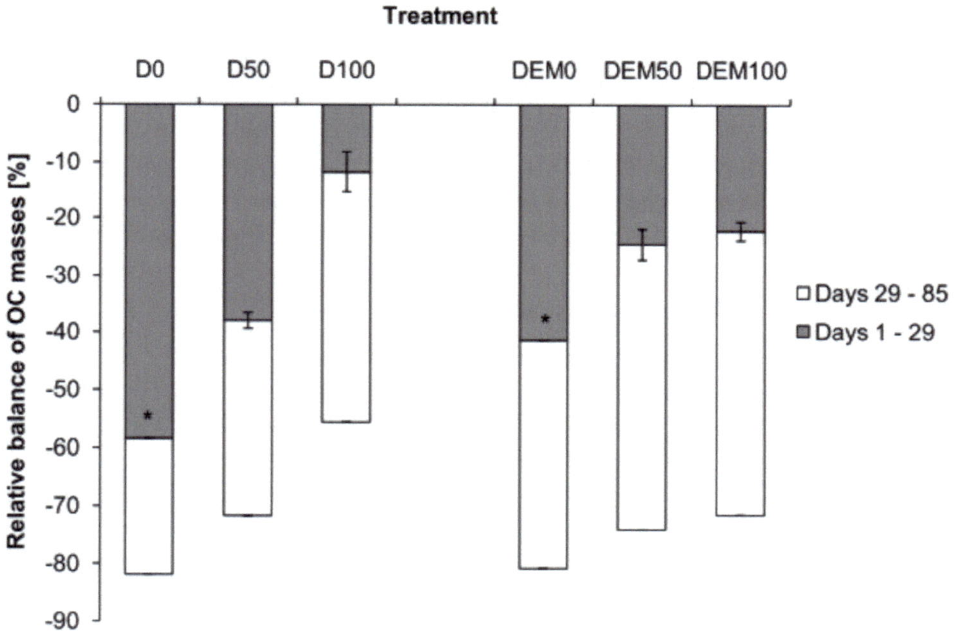

Fig. 5. Relative mass balance of organic carbon (OC) during fermentation (Days 1 - 29) and composting (Days 9 - 85; means ± standard error; significant differences between 'Days1-29' OC losses of D0 and DEM0 (see asterisks), $p < 0.05$, n=3, tested with a Student's t-test). D0, D50 and D100 are composted materials at 0, 50 and 100 kg biochar addition per ton of composted materials, respectively, For DEM 0, 50 and 100 it was the same approach but during the first 29 days of the experiment, these piles were incubated with effective microorganisms under anaerobic conditions (Erben, 2011).

Fermentation-induced negative priming of fresh OM could indeed be observed, but only as a temporary effect which was reversed during subsequent composting. Thus, fermentation did not result in an enhanced stabilization of compost. A distinct effect of the EM preparation could not be identified (Erben, 2011). Nevertheless, benefits of fermentation in OM treatment and for soil application remain to be assessed.

4.2 C sequestration and priming as function of biochar amount

Biochar could cause a positive priming effect due to its high surface area providing habitat for microorganisms and due to input of partly labile C substrate (condensates). On the other hand, biochar is a stable compound which could stabilize labile compost OM thus providing a negative priming effect.

Composting of biochar could be successfully conducted over a wide biochar / organic material ratio covering up to 50% biochar by weight. During composting, a relative enrichment of biochar was observed which is obvious as biochar is much more stable than organic waste materials (Erben, 2011). However, biochar caused a significant positive priming effect on non-biochar composting materials at low (up to 1 weight%) biochar concentrations (Erben 2011) while at high (up to 50 weight%) biochar concentrations a significantly negative priming effect could be observed (Erben, 2011; Fig. 5). Therefore, a synergistic benefit for overall C sequestration could be observed when biochar was composted together with organic waste material (Erben, 2011). Further co-benefits might arise for soil microbial biomass and community structure composition and for biochar surface oxidation which still has to be proven scientifically.

Combining biochar addition and fermentation resulted in negative priming (Fig. 5), but the effect was weaker here than that of non-fermented treatments and hence ascribed rather to biochar alone than to its reinforcement of fermentation-induced negative priming.

4.3 Synergisms for soil fertility and plant growth

Combination of biochar with inorganic and organic fertilizers is clearly advantageous over the sole biochar or fertilizer amendments (Fig. 6). Plant growth significantly increased after biochar addition. Although pure compost application showed highest absolute yield during two growth periods, biochar compost mixture revealed highest relative performance. It should be mentioned here that biochar compost mixture received only 50% of pure biochar and 50% of pure compost treatments, thus providing evidence for biochar compost synergism (Fig. 6). In addition, it can be expected that in the long-term, compost will be mineralized more quickly than biochar or compost biochar mixtures. Mineral fertilizer retention was significantly more efficient when biochar was present although biochar did not increase cation exchange capacity at least after the first harvest (Schulz & Glaser, 2011). In comparison to mere mineral fertilizer there were clear advantages of plant growth and soil quality of the biochar-amended soils, especially when combining fertilizer (both inorganic and organic) with biochar. Therefore, optimization of biochar compost systems will be discussed in the following.

In a greenhouse experiment on a sandy soil under temperate climate conditions, plant growth (and thus soil fertility) generally increased with increasing amendment of biochar-compost (Fig. 7). This effect is more pronounced in a (nutrient-poor) sandy soil compared to a loamy soil (Fig. 7). It is interesting to note however, that at individual application rates, a synergistic effect of higher biochar application is obvious in the sandy soil (Fig. 7). This is even more interesting as biochar application rates were generally low with a maximum of 10 kg biochar per ton of compost material.

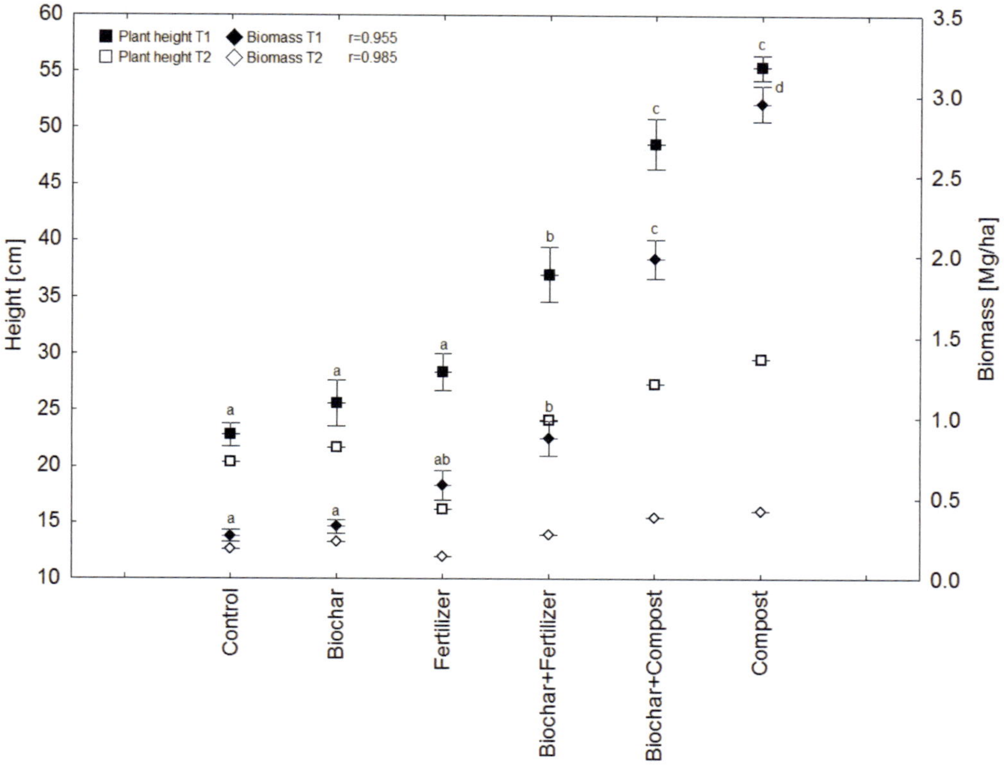

Fig. 6. Crop (oats, *Avena sativa*) response of two consecutive harvests on a sandy soil amended with different materials. Treatments comprised control (only water), mineral fertilizer (111.5 kg N ha^{-1}, 111.5 kg P ha^{-1} and 82.9 kg K ha^{-1}), compost (5% by weight), biochar (5% by weight) and combinations of biochar (5% by weight) plus mineral fertilizer (111.5 kg N ha^{-1}, 111.5 kg P ha^{-1} and 82.9 kg K ha^{-1}) and biochar (2.5% by weight) plus compost (2.5% by weight) (Schulz & Glaser, 2011).

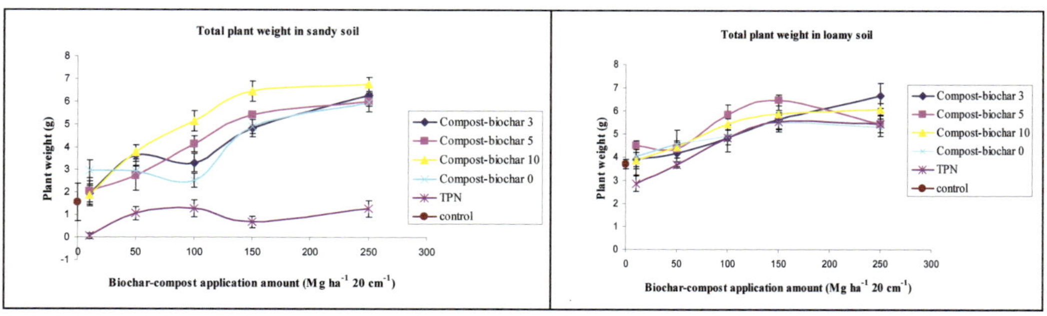

Fig. 7. Crop (oats, *Avena sativa*) response on a sandy (left) and loamy (right) soil with increasing biochar-compost amendments (x axis) at low biochar additions (3, 5 and 10 kg per ton of compost, different symbols) compared to control soil (without amendments) and a commercial biochar-containing product (TPN) (Schulz and Glaser, unpublished).

When looking at high biochar amounts, crop (oats, *Avena sativa*) yield significantly increased with increasing amounts of biochar and compost amendments, both for sandy (Fig. 8 left) and loamy soils (Fig. 8 right). However, in both cases, plant growth response was higher for biochar than for compost (sand: plant weight = 2.490 + 0.00676 compost + 0.0400 biochar, loam: plant weight = 4.088 + 0.0144 compost + 0.0349 biochar).

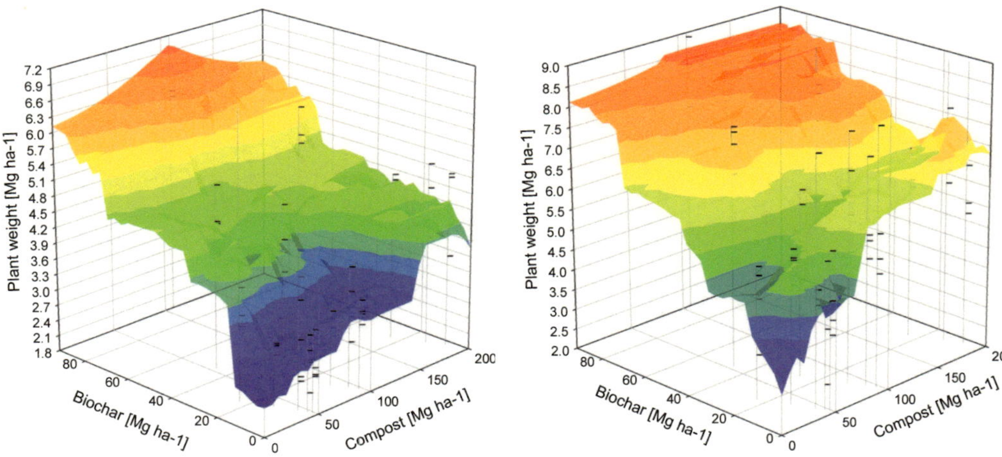

Fig. 8. Crop (oats, *Avena sativa*) response on a sandy (left) and loamy (right) soil with increasing biochar-compost amendments at high biochar additions (Schulz & Glaser, unpbulished).

4.4 Can combined biochar compost processing contribute to optimized material flow management?

By taking into account that *terra preta* formation was originally induced by human activity relying on the combined incorporation and biological transformation of charred stable OM on the one hand and nutrient-rich, organic feedstocks on the other hand (Fig. 3), it seems obvious that *terra preta* genesis can be understood as a sustainable and optimized management of natural resources. However, *terra preta* soils do not normally occur under conditions in which just compost or mulching material have been applied. Therefore, the addition of biochar can be recognized as a key factor for the reproduction of Terra preta similar substrates (chapter 3.1). However, the sole addition of charred biomass does also not result into the formation of *terra preta* soils. Thus, nutrient incorporation and microbial activity can be specified as further key factors.

In this respect, it seems to be a promising approach to combine the existing scientific knowledge about ancient *terra preta* genesis with modern composting technology to promote positive, synergestic effects for an efficient and optimized management of natural resources including 'organic wastes' to create humus and nutrient-rich substrates with beneficial effects for soil amelioration, carbon sequestration and sustainable land use systems. Fig. 9 gives a synthesis of the information about composting and biochar application and their beneficial effects hitherto presented in this review to show options for a sustainable material flow management.

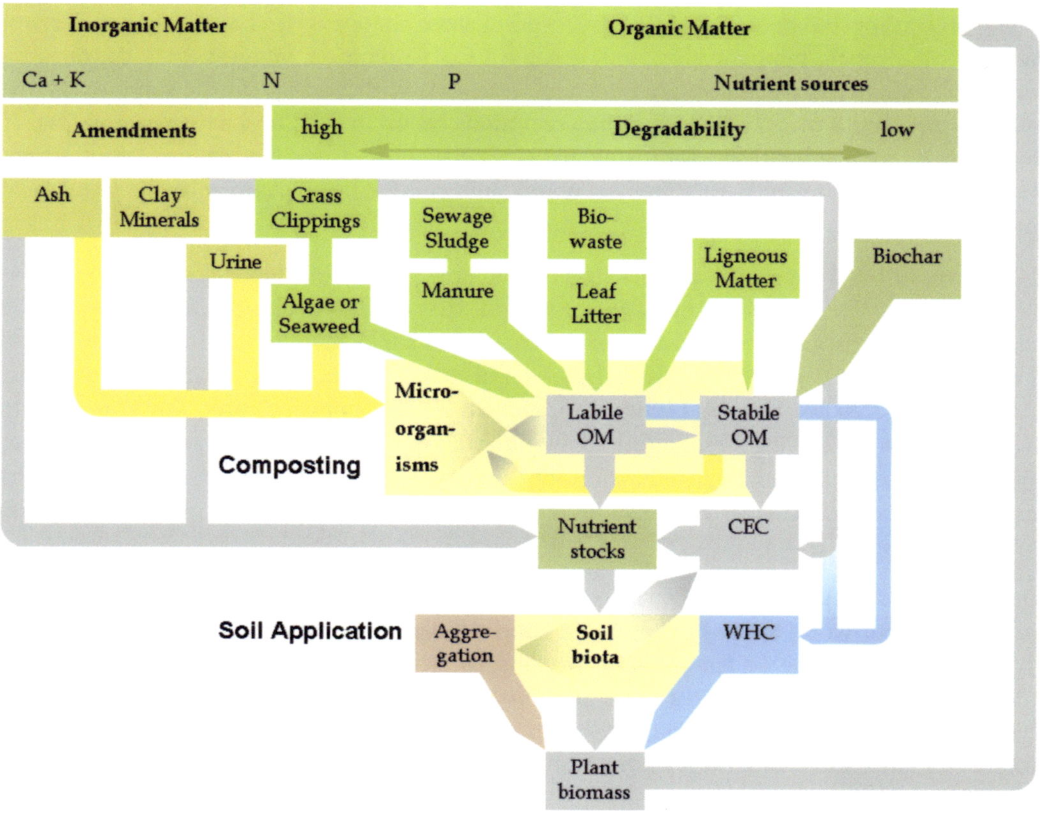

Fig. 9. Sustainable management of natural resources by combining biochar with organic and inorganic wastes in compost processing (based on Glaser & Birk, 2011).

Based on the model of *terra preta* genesis (Glaser & Birk, 2011) various organic and inorganic feedstocks are mixed for composting providing different nutrients resources. Ideally, their physico-chemical properties should complete each other promoting an appreciable C/N ratio, water content, aeration, nutrient composition etc. of the initial compost pile. Besides their nutrient level, the used organic input materials can be characterized by their biological degradability and their contribution to different carbon pools. N-rich feedstocks such as grass clippings are easily decomposable particularly contributing to the labile OM pool which is used as an easy available food source of microorganisms and thus providing optimum conditions for a rapid rotting process. In contrast, ligneous materials are characterized by a lower degradability due to their higher lignin content partially contributing to the stable OM pool which has beneficial long-term effects for soil amelioration, carbon sequestration (Fig. 1) as well as humus reproduction (Table 1). The most recalcitrant material towards biological degradation is represented by biochar contributing at most to the stable OM pool of substrate mixtures. During subsequent aerobic decomposition OM getting stabilized resulting in an increase of stable C content. According to Yoshizawa et al. (2005) biochar promotes this rotting process due to its functions as a matrix for the involved aerobic microorganisms probably increasing decomposition speed. An co-composting experiment with poultry litter and biochar applied by Steiner et al. (2010)

seems to confirm the accuracy of this assumption since changes in pH and moisture content with greater peak temperatures and greater CO_2 respiration suggest that composting process was more rapid if poultry litter was amended with biochar. In the same study the authors detected a reduction of ammonia emissions by up to 64 % and a decrease of total N losses by up to 52% if poultry litter was mixed with biochar. These observations support the hypothesis of higher nutrient retention ability induced by biochar amendment previously mentioned in this review.

Furthermore by the proliferation of microorganisms on the biochar backbone as well as between its pores, Yoshizawa et al. (2005) suggest that biochar properties are influenced by biological processes. Especially slow oxidation of biochar over time has been suggested to produce carboxylic groups on the edges of the aromatic backbone, increasing the CEC (Glaser et al., 2000). Due to higher temperature during compost processing, especially during thermophilic stage, biological activity as well as chemical reaction rate is increased, probably accelerating the partial oxidation and formation of functional groups of the amended biochar material but also interaction with labile OM and with minerals is favoured.

Besides the importance of biochar incorporation, additional amendments like clay minerals can add further value to the final compost product, e.g. by promoting an enhanced CEC or WHC due to their high adsorption or swelling capacity. Furthermore, their incorporation into organic substrates promotes the formation of organo-mineral complexes initiated by the biological activity of soil fauna after subsequent soil application. This aspect seems important since SOM in *terra preta* is stabilized by interaction with soil minerals (Glaser et al., 2003).

Other amendments like ash, excrements or urine contribute to the nutrients stock of the final composting product and can enhance microbial activity by their nutrient supply (Glaser & Birk 2011). According to Arroyo-Kalin et al. (2009) and Woods (2003), ash may have been a significant input material into *terra preta,* too. Furthermore for providing adequate moisture conditions during composting urine can be added instead of water for preventing the dehydration of composting piles while adding nutrients at the same time.

After compost maturation, the final compost substrate can be beneficially applied to soils. In this respect, the soil biota contribute to a further transformation of the applied material and provide essential ecological services, for instance by promoting aggregation and further OM stabilization. By enhancing the specific biological, physical and chemical properties of soils amended with the biochar composting substrates, plant growth is generally promoted.

5. Conclusions

Our review clearly demonstrated beneficial effects of compost for ecosystem services. In addition, it is a promising tool for sustainable management of natural resources (soils, organic 'waste'. Especially two of the major problems of modern society (anthropogenic greenhouse effect and desertification) could be coped with proper compost technologies. However, as compost has only a moderate SOM reproduction potential, strategies for further optimization are required. These could be applying the *terra preta* concept, especially the integration of biochar into management of natural resources. Recent studies provide optimism for synergistic effects of compost and biochar technologies for ecosystem services and for sustainable management of natural resources including 'organic wastes'.

6. References

Ahmad, Z.; Yamamoto, S. & Honna, T. (2008). Leachability and Phytoavailability of Nitrogen, Phosphorus, and Potassium from Different Bio-composts under Chloride- and Sulfate-Dominated Irrigation Water, *J. Environ. Qual.*, Vol. 37, pp. 1288–1298.

Alluvione, F.; Bertora, C.; Zavattaro, L. & Grignani, C. (2010). Nitrous Oxide and Carbon Dioxide Emissions Following Green Manure and Compost Fertilization in Corn, *Soil Sci. Soc. Am. J.*, Vol. 74, No. 2, pp. 384–395.

Amlinger, F.; Peyr, S.; Geszti, J.; Dreher, P.; Karlheinz, W. & Nortcliff, S. (2007). *Beneficial effects of compost application on fertility and productivity of soils. Literature Study*, Federal Ministry for Agriculture and Forestry, Environment and Water Management, Austria, Retrieved from www.umweltnet.at/filemanager/download/20558/

Annabi, M.; Houot, S.; Francou, C.; Poitrenaud, M. & Le Bissonnais, Y. (2007). Soil Aggregate Stability Improvement with Urban Composts of Different Maturities, *Soil Sci. Soc. Am. J.*, Vol. 71, No. 2, pp. 413–423.

Arroyo-Kalin M.; Neves E. G. & Woods W. I. (2009). Anthropogenic Dark Earths of the Central Amazon region: remarks on their evolution and polygenetic composition, In: *Amazonian Dark Earths: Wim Sombroek's Vision*, Woods et al. (Eds.), pp. 99–125, Springer, Berlin.

Atkinson, C.J.; Fitzgerald, J.D. & Hipps, N.A. (2010). Potential mechanisms for achieving agricultural benefits from biochar application to temperate soils: a review, *Plant Soil*, Vol. 337, pp. 1-18.

Badr EL-Din, S.M.S.; Attia, M. & Abo-Sedera, S. A. (2000). Field assessment of composts produced by highly effective cellulolytic microorganisms, *Biology and Fertility of Soils*, Vol. 32, pp. 35–40.

Bar-Tal, A.; Yermiyahu, U.; Beraud, J.; Keinan, M.; Rosenberg, R.; Zohar, D.; Rosen, V. & Fine, P. (2004). Nitrogen, phosphorus, and potassium uptake by wheat and their distribution in soil following successive, annual compost applications, *J. Environ. Qual.*, Vol. 33, pp. 1855–1865.

Becker, J.; Hartmann, R. & Hubrich, J. (1995). *Das Modell des standortgerechten Kompostes. Entwicklung des Modells und dessen Anwendung für drei Teilräume des Bremer Umlandes*, Univ.-Buchh., ISBN 978-3-88722-338-0, Bremen

Beck-Friis, B.; Smårs, S.; Jönsson, H. & Kirchmann, H. (2001). Gaseous emissions of carbon dioxide, ammonia, and nitrous oxide from organic household waste in a compost reactor under different temperature regimes, J. Agric. Eng. Res., Vol. 78, pp. 423–430.

Beraud, J.; Fine, P.; Yermiyahu, U.; Keinan, M.; Rosenberg, R.; Hadas, A. & Bar-Tal, A. (2005). Modeling carbon and nitrogen transformations for adjustment of compost application with nitrogen uptake by wheat, *J. Environ. Qual.*, Vol. 34, pp. 664–675.

Bidlingmaier, W. & Gottschall, R. (2000). *Biologische Abfallverwertung*, Ulmer, ISBN 3800132087, Stuttgart (Hohenheim)

Birk, J. J.; Steiner, C.; Teixiera, W. C.; Zech, W. & Glaser, B. (2009). Microbial Response to Charcoal Amendments and Fertilization of a Highly Weathered Tropical Soil, In: *Amazonian Dark Earths: Wim Sombroek's Vision,* Woods, W. I.; Teixeira, W. G.; Lehmann, J.; Steiner, C.; WinklerPrins A. and Rebellato, L. (Eds.), pp. 309-324. Springer, ISBN 978-1-4020-9030-1

Bischoff, R. (1988). Auswirkungen langjähriger differenzierter organischer Düngung auf Ertrag und Bodenparameter, In: *Abfallstoffe als Dünger. Möglichkeiten und Grenzen : Vorträge zum Generalthema des 99. VDLUFA-Kongresses 14. - 19.9.1987 in Koblenz,* VDLUFA-Schriftenreihe 23, Zarges, H. (Ed.), pp. 451-466, VDLUFA-Verlag, ISBN 3922712282, Darmstadt

Blume, H.-P. (1989). Organische Substanz, In: *Lehrbuch der Bodenkunde*, Scheffer, F. & Schachtschabel, P. (Eds.), 12. Aufl., neu bearb., Enke, ISBN 3432847726, Stuttgart.

Buchmann, I. (1972). Nachwirkungen der Müllkompostanwendung auf die bodenphysikalischen Eigenschaften, *Landwirtschaftliche Forschung,* 28 (1), pp. 358-362.

Bundesgütegemeinschaft Kompost e. V. [BGK] & Bundesforschungsanstalt für Landwirtschaft [FAL] 2006). *Organische Düngung. Grundlagen der guten fachlichen Praxis.* (3rd Edition), Bundesgütegemeinschaft Kompost e.V. Köln, Retrieved from www.kompost.de/fileadmin/docs/shop/Anwendungsempfehlungen/Organische_Duengung_Auflage3.pdf

Carter, M. R.; Sanderson, J. B. & MacLeod, J. A. (2004). Influence of compost on the physical properties and organic matter fractions of a fine sandy loam throughout the cycle of a potato rotation. *Canadian Journal of Soil Science,* 84, pp. 211-218.

Cheng, C. H.; Lehmann, J.; Thies, J. E. & Burton, S. D. (2008). Stability of black carbon in soils across a climatic gradient. *Journal of Geophysical Research-Biogeosciences,* 113.

Dalal, R.C.; Gibson, I.R. & Menzies, N.W. (2009). Nitrous oxide emission from feedlot manure and green waste compost applied to Vertisols, *Biology and Fertility of Soils,* Vol. 45, pp. 809–819.

Diacono, M. & Montemurro, F. (2010). Long-term effects of organic amendments on soil fertility. A review, Agron. Sustain. Dev., vol. 30, No.2, pp. 401–422, DOI 10.1051/agro/2009040

Diez, T. & Kraus, M. (1997). Wirkung langjähriger Kompostdüngung auf Pflanzenertrag und Bodenfruchtbarkeit, *Agribiological Research,* 50, pp. 78 - 84.

Eklind, Y.; Beck-Friis, B.; Bengtsson, S.; Ejlertsson, J.; Kirchmann, H.; Mathisen, B.; Nordkvist, E.; Sonesson, U.; Svensson, B.H. & Torstensson, L. (1997). Chemical characterization of source-separated organic household wastes, *Swed. J. Agric. Res.,* Vol. 27, pp. 167–178.

Erben, G. A. (2011). Carbon dynamics and stability of biochar compost. An evaluation of three successive composting experiments, Bachelor Thesis, University of Bayreuth, Bayreuth.

Evanylo, G.; Sherony, C.; Spargo, J.; Starner, D.; Brosius, M. & Haering, K. (2008). Soil and water environmental effects of fertilizer-, manure-, and compost-based fertility practices in an organic vegetable cropping system. *Agriculture, Ecosystems & Environment* 127, 50-58.

Flaig, W. (1968). Einwirkung von organischen Bodenbestandteilen auf das Pflanzenwachstum, *Landwirtsch. Forschung*, Vol. 21, pp. 103–127.

Fox, R. (1986). Ergebnisse aus einem Abdeckungsversuch in Steillagen. *Rebe und Wein*, 39, pp. 357- 360.

Fricke, K.; Turk, T. & Vogtmann, H. (1990). Grundlagen der Kompostierung. Berlin: EF-Verl. für Energie- und Umwelttechnik (Technik, Wirtschaft, Umweltschutz).

Gerber, H. (2010). Dezentrale CO2-negative energetische Biomasseverwertung mit dem PYREG-Verfahren, *Proceedings of „Biokohle Workshop" IFZ Gießen*, University Gießen, 23.-24. Feb. 2010, Retrieved from www.uni-giessen.de/cms/fbz/fb08/biologie/pflanzenoek/forschung/workshop/copy_of_workshop/gerber/view

Glaser, B. (2007). Prehistorically modified soils of central Amazonia: a model for sustainable agriculture in the twenty-first century. *Philosophical Transactions of the Royal Society B-Biological Sciences* 362, 187-196.

Glaser, B. & Birk, J. J. (2011). State of the scientific knowledge on properties and genesis of Anthropogenic Dark Earths in Central Amazonia (terra preta de Índio). *Geochimica Et Cosmochimica Acta* doi:10.1016/j.gca.2010.11.029.

Glaser, B.; Balashov, E.; Haumaier, L.; Guggenberger, G. & Zech, W. (2000). Black carbon in density fractions of anthropogenic soils of the Brazilian Amazon region. *Organic Geochemistry* 31, 669-678.

Glaser, B.; Guggenberger, G.; Zech, W. & de Lourdes Ruivo, M. (2003). Soil organic matter stability in Amazon Dark Earth. In: *Amazonian Dark Earths: Origin, Properties, Management*, Lehmann et al. (Ed.), Kluwer Academic Publishers, pp. 141–158, Dodrecht.

Glaser, B.; Haumaier, L.; Guggenberger, G. & Zech, W. (2001). The Terra Preta phenomenon: a model for sustainable agriculture in the humid tropics. *Naturwissenschaften* 88, 37-41.

Glaser, B.; Lehmann, J. & Zech, W. (2002). Ameliorating physical and chemical properties of highly weathered soils in the tropics with charcoal - a review. *Biology and Fertility of Soils* 35, 219-230.

Gottschall, R. (1984). *Kompostierung. Optimale Aufbereitung und Verwendung organischer Materialien im ökologischen Landbau*, Müller, ISBN 3-7880-9687-X, Karlsruhe.

Hadar, Y.; Mandelbaum, R. & Gorodecki, B. (1992). Biological control of soilborne plant pathogens by suppressive compost.

Harms, H. (1983). Phenolstoffwechsel von Pflanzen in Abhängigkeit von Stickstofform und -angebot. *Landwirtsch. Forschung*, Vol. 36, pp. 9–17.

Hartmann, R. (2003). Studien zur standortgerechten Kompostanwendung auf drei pedologisch unterschiedlichen, landwirtschaftlich genutzten Flächen der Wildesauer Geest, Niedersachsen. Bremen, Germany, Universität Bremen, Institut für Geographie.

Haug, R.T. (1993). The practical handbook of compost engineering, LEWIS PUBLISHERS, ISBN 0-87371-373-7, Boca Raton (Florida)

Higa, T.; Wididana, G.N. (1991). The concept and theories of effective microorganisms. *Proceedings of the first international conference on Kyusei nature farming,* US Department of Agriculture, Washington DC, www.infrc.or.jp/english/KNF_Data_Base_Web/PDF%20KNF%20Conf%20Data/C 1-5-015.pdf.

Hoitink, H.A.J. (1980). Composted bark, a light weight growth medium with fungicidal properties, *Plant. Dis. 64,* pp. 142-147

Hoitink, H.A.J. & Fahy, P.C. (1986). basis for the control of soilborne plant-pathogens with composts. In: *Annu Rev Phytopathol 24,* S. 93–114.

Holmgren, G.G.S., Meyer, M.W., Chaney, R.L., Daniel, D.B. (1993). Cadmium, lead, zinc, copper and nickel in agricultural soils of the United States of America. Journal of Environmental Quality, 22: 335-348.

Hudson, B. D. (1994). Soil organic matter and available water capacity. J. of soil & water conservation, 49, 189 -194.

Kahle, P. & Belau, L. (1998). Modellversuche zur Prüfung der Verwertungsmöglichkeiten von Bioabfallkompost in der Landwirtschaft, *Agribiological Research,* 51, pp. 193 - 200.

Kehres, B. (1992). Begrenzung der Kompostausbringung durch Nährstoffe. In: *Gütesicherung und Vermarktung von Bioabfallkompost,* Wiemer, K. & Kern, M. (Eds.), Abfall - Wirtschaft 9, pp. 395-408, Witzenhausen-Institut für Abfall, Umwelt und Energie, ISBN 978-3-928673-02-0, Witzenhausen

Kimetu, J.M.; Lehmann, J.; Ngoze, S.O.; Mugendi, D.N.; Kinyangi, J.M.; Riha, S.; Verchot, L.; Recha, J.W. & Pell, A.N. (2008). Reversibility of soil productivity decline with organic matter of differing quality along a degradation gradient, *Ecosystems,* Vol. 11, pp. 726–739.

Kögel-Knabner, I.; Leifeld, J. & Siebert, S. (1996). Humifizierungsprozesse von Kompost nach Ausbringung auf den Boden. In: *Neue Techniken der Kompostierung. Kompostanwendung, Hygiene, Schadstoffabbau, Vermarktung, Abluftbehandlung - Dokumentation,* Stegmann, R. (Ed.), Hamburger Berichte 11, pp. 73-87, Economica Verlag, ISBN 3-87081-196-X, Bonn

Kong, A. Y. Y.; Six, J.; Bryant, D. C.; Denison, R. F. & van Kessel, C. (2005). The relationship between carbon input, aggregation, and soil organic carbon stabilization in sustainable cropping systems. *Soil Science Society of America Journal* 69, 1078-1085.

Krieter, M. (1980), Bodenerosionen in Rheinhessischen Weinbergen, 3. Teil, *ifoam,* Nr. 35, pp. 10-13.

Kuzyakov, Y.; Subbotina, I.; Chen, H. Q.; Bogomolova, I. & Xu, X. L. (2009). Black carbon decomposition and incorporation into soil microbial biomass estimated by C-14 labeling. *Soil Biology & Biochemistry* 41, 210-219.

Lal, R. (2009). Soils and food sufficiency. A review. *Agronomy for Sustainable Development* 29, 113-133.

Larney, F.J.; Olson, A.F.; Miller, J.J.; DeMaere, P.R.; Zvomuya, F. & McAllister, T.A. (2008). Physical and chemical changes during composting of wood chip-bedded and straw-bedded beef cattle feedlot manure, *Journal of Environmental Quality,* Vol. 37, pp. 725–735.

Lechner, P.; Linzner, R.; Mostbauer, P.; Binner, E.; Smidt, E. (2005), Klimarelevanz der Kompostierung unter Berücksichtigung der Verfahrenstechnik und Kompostanwendung (KliKo), *Endbericht im Auftrag der MA 48*, Universität für Bodenkultur Wien, Department für Wasser - Atmosphäre - Umwelt, Retrieved from www.boku.ac.at/TCG/rol/KliKo_Endbericht.pdf

Lee, J. W.; Hawkins, B.; Day, D. M. & Reicosky, D. C. (2010). Sustainability: the capacity of smokeless biomass pyrolysis for energy production, global carbon capture and sequestration. *Energy & Environmental Science* 3, 1695–1705.

Lehmann, J.; Skjemstad, J. & Sohi, S. (2008). Australian climate-carbon cycle feedback reduced by soil black carbon. *Nature Geoscience* 1, 832–835.

Leita, L.; Fornasier, F.; Mondini, C. & Cantone, P. (2003). Organic matter evolution and availability of metals during composting of MSW, In: *Applying Compost - Benefits and Needs. Seminar Proceedings Brussels, 22 - 23 November 2001*, Federal Ministry of Agriculture, Forestry, Environment and Water Management, Austria, and European Communities, pp. 201-206, ISBN 3-902 338-26-1, Brussels, Retrieved from http://ec.europa.eu/environment/waste/compost/seminar.htm

Lenzen, P. (1989): Untersuchungsergebnisse zur Verwendung von Müllkompost und Müllklärschlammkompost zur Bodenverbesserung und Bodenherstellung, III, Bestandsentwicklung, Biomassebildung und Nährstoffentzug, *Z. f. Vegetationstechnik im Landschafts- und Sportstättenbau 12*, pp. 81-96.

Liang, B.; Lehmann, J.; Solomon, D.; Sohi, S.; Thies, J. E.; Skjemstad, J. O.; Luizao, F. J.; Engelhard, M. H.; Neves, E. G. & Wirick, S. (2008). Stability of biomass-derived black carbon in soils. *Geochimica Et Cosmochimica Acta* 72, 6069-6078.

Liu, B.; Gumpertz, M. L.; Hu, S. & Ristaino, J. B. (2007). Long-term effects of organic and synthetic soil fertility amendments on soil microbial communities and the development of southern blight. *Soil Biology and Biochemistry* 39, 2302-2316.

Löbbert, M. & Reloe, H. (1991). Verfahren der Ausbringung aufbereiteter organischer Reststoffe zur Verminderung der Erosion in Reihenkulturen (Mais), *Arbeiten aus dem Institut für Landtechnik der Universität Bonn*, Heft 7.

Lützow, M. von; Kögel Knabner, I.; Ludwig, B.; Matzner, E.; Flessa, H.; Ekschmitt, K.; Guggenberger, G.; Marschner, B. & Kalbitz, K. (2008). Stabilization mechanisms of organic matter in four temperate soils: Development and application of a conceptual model. *Journal of Plant Nutrition and Soil Science*, Vol. 171, No. 1, pp. 111–124.

Marschner, B. & Flessa, H. (2006): Stabilization of organic matter in temperate soils: mechanisms and their relevance under different soil conditions - a review. *European Journal of Soil Science*, Vol. 57, pp. 426–445.

Neklyudov, A. D.; Fedotov, G. N. & Ivankin, A. N. (2006): Aerobic Processing of Organic Waste into Composts. In: *Applied Biochemistry & Microbiology 42 (4)*, pp. 341–353.

Nelson, E. B. & Hoitink, H. A. (1983). The role of microorganisms in the suppression of Rhizoctonia solani in container media amended with composted hardwood bark. In: *Phytopathology 73 (2)*, S. 274–278.

Nguyen B.T., Lehmann J., Hockaday W.C., Joseph S., Masiello C. (2010). Temperature sensitivity of black carbon decomposition and oxidation, *Environ. Sci. Technol.*, Vol. 44, No. 9, pp. 3324-3331.

Nishio, M. (1996). Microbial Fertilizers in Japan. Retrieved from: www.agnet.org/library/eb/430

Ogawa, M.; Yambe Y. & Sugiura, G. (1983). Effects of charcoal on the root nodule formation and VA mycorrhiza formation of soy bean. *Proceedings of The Third International Mycological Congress (IMC3)*, Abstract, p. 578

Ogawa, M. (1994). Symbiosis of people and nature in the tropics, In: *Farming Japan 28* (5), pp. 10–30.

Ouedraogo, E.; Mando, A. & Zombre, N. P. (2001). Use of compost to improve soil properties and crop productivity under low input agricultural system in West Africa. *Agriculture Ecosystems & Environment* 84, 259-266.

Smith, J.L. & Collins, H.P. (2007). Composting. In: *Soil Microbiology, Ecology, and Biochemistry* (3rd edition), Paul, E.A., (Ed)., pp. 483-486. Academic Press, ISBN 0125468075, 9780125468077, Burlington

Pietikäinen, J.; Kiikkila, O. & Fritze, H. (2000). Charcoal as a habitat for microbes and its effect on the microbial community of the underlying humus. *Oikos* 89, 231-242.

Richter, G. (1979). Bodenerosion in Reblagen des Moselgebietes, Ergebnisse quantitativer Untersuchungen, 1974-1977: Forschungsstelle Bodenerosion d. Univ. Trier.

Roberts, K. G.; Gloy, B. A.; Joseph, S.; Scott, N. R. & Lehmann, J. (2010). Life Cycle Assessment of Biochar Systems: Estimating the Energetic, Economic, and Climate Change Potential. *Environmental Science & Technology* 44, 827-833.

Reinhold, J. (2009). Betrachtungen von Kompostqualitäten und deren Ausgangsmaterialien (Biomasse), Proceedings of *Schließung von Stoffkreisläufen – Kohlenstoffkreislauf – Veranstaltung der Kommission Bodenschutz und des Umweltbundesamtes*, Dessau, 19.-20. Nov. 2009, posting

Rondon, M. A.; Lehmann, J.; Ramirez, J. & Hurtado, M. (2007). Biological nitrogen fixation by common beans (Phaseolus vulgaris L.) increases with bio-char additions. *Biology and Fertility of Soils* 43, 699-708.

Sahin, H. (1989). Auswirkungen des langjährigen Einsatzes von Müllkompost auf den Gehalt an organischer Substanz, die Regenwurmaktivität, die Bodenatmung sowie die Aggregatstabilität und die Porengrößenverteilung, *Mitteilgn. Dtsch. Bodenkundl. Gesellsch.*, Vol. 59, No. II, pp. 1125–1130.

Saito, M. & Marumoto, T. (2002). Inoculation with arbuscular mycorrhizal fungi: the status quo in Japan and the future prospects. *Plant and Soil* 244, 273-279.

Schimmelpfennig, S. & Glaser, B. (2012). Material properties of biochars from different feedstock material and different processes. *Journal of Environmental Quality* in press. doi:10.2134/jeq2011.0146

Schulz, H. & Glaser, B. (2011). Biochar - benefits for soil and plants as compared with conventional soil amendments. *Journal of Plant Nutrition and Soil Science* in press.

Seiberth, W. & Kick, H. (1969): Ein zwölfjähriger Freilandversuch zur Wirkung von Müll- und Klärschlammkomposten, Stallmist und Stroh auf Ertrag, Nährstoff- und Humusgehalt des Bodens, *Landw. Forschung* 23, pp. 13- 22.

Smith J.L.; Collins, H.P.; Bailey V.L. (2010). The effect of young biochar on soil respiration, *Soil Biology and Biochemistry*, Vol. 42, Issue 12, pp. 2345-2347

Söchtig, H. (1964): Beeinflussung des Stoffwechsels der Pflanzen durch Humus und seine Bestandteile und die Auswirkung auf Wachstum und Ertrag. *Landbauforschung Völkenrode*, Vol. 14, pp. 9–16.

Stamatiadis, S.; Werner, M. & Buchanan, M. (1999). Field assessment of soil quality as affected by compost and fertilizer application in a broccoli field, *Applied Soil Ecology*, 12, pp. 217 - 225.

Steiner, C.; Das, K.C.; Melear, N.; Lakly, D. (2010). Reducing Nitrogen Loss during Poultry Litter Composting Using Biochar. *Journal of Environmental Quality*, Vol. 39, pp. 1236–1242.

Steiner, C; Melear, N.; Harris, K. & Das, K.C. (2011). Biochar as bulking agent for poultry litter composting, *Carbon Management*, June 2011, Vol. 2, No. 3, pp. 227-230

Steiner, C.; Teixeira, W. & Zech, W. (2004). Slash and char - an alternative to slash and burn practiced in the Amazon Basin. *In* "Amazonian Dark Earths" (B. Glaser and W. Woods, eds.), pp. 182-193. Springer, Heidelberg.

Steiner, C.; Glaser, B.; Teixeira, W. G.; Lehmann, J.; Blum, W. E. H. & Zech, W. (2008). Nitrogen retention and plant uptake on a highly weathered central Amazonian Ferralsol amended with compost and charcoal. *Journal of Plant Nutrition and Soil Science-Zeitschrift Fur Pflanzenernahrung Und Bodenkunde* 171, 893-899.

Stöppler-Zimmer, H.; Gottschall, R. & Gallenkemper, B. (1993). Anforderungen an Qualität und Anwendung von Bio- und Grünkomposten, *Schriftenreihe des Arbeitskreises für die Nutzbarmachung von Siedlungsabfällen (ANS) e. V.*, Heft 25, Studie im BMFT-Verbundvorhaben „Neue Techniken der Kompostierung", 1st Edition, Kurz & Co. Druckerei + Reprografie GmbH, Stuttgart

Strauss, P. (2003). Runoff, soil erosion and related physical properties after 7 years of compost application. In: *Applying Compost - Benefits and Needs. Seminar Proceedings Brussels, 22 - 23 November 2001*, Federal Ministry of Agriculture, Forestry, Environment and Water Management, Austria, and European Communities, pp. 219-224, ISBN 3-902 338-26-1, Brussels, Retrieved from http://ec.europa.eu/environment/waste/compost/seminar.htm

Tejada, M.; Garcia, C.; Gonzalez, J. L. & Hernandez, M. T. (2006). Use of organic amendment as a strategy for saline soil remediation: Influence on the physical, chemical and biological properties of soil. *Soil Biology & Biochemistry* 38, 1413-1421.

Termorshuizen, A. J.; van Rijn, E. & Blok, W. J. (2005). Phytosanitary risk assessment of composts. *Compost Science & Utilization* 13, 108-115.

Thies J., Rillig M.C. (2009). Characteristics of biochar: Biological properties. In: *Biochar for environmental management: Science and technology*, Lehmann J & Joseph S. (Eds)., pp. 85-105, Earthscan, London

Tisdall, J. M. & Oades, J. M. (1982). Organic matter and water stable aggregates in soils. *Journal of Soil Science* 33, 141-161.

Tryon, E. H. (1948). Effect of Charcoal on Certain Physical, Chemical, and Biological Properties of Forest Soils. *Ecological Monographs* 18, 81-115.

United States Environmental Protection Agency (1998). *An Analysis of Composting as an Environmental Remediation Technology*, Solid Waste and Emergency Response, EPA530-R-98-008, Washington DC

Vaz-Moreira, I.; Silva, M.E.; Manaia, C.M. & Nunes, O.C. (2007). Diversity of bacterial isolates from commercial and homemade composts, *Microbial Ecology*, Vol. 55, No. 4, pp. 714–722.

Verheijen, F. G. A.; Jeffery, S.; Bastos, A. C.; van der Velde, M. & Diafas, I. (2009). "Biochar Application to Soils - A Critical Scientific Review of Effects on Soil Properties, Processes and Functions," Luxembourg.

Vogtmann, H.; Kehres, B.; Gottschall, R. & Meier-Ploeger, A. (1991). Untersuchungen zur Kompostverwertung in Landwirtschaft und Gartenbau, In: *Bioabfallkompostierung - flächendeckende Einführung, Abfall-Wirtschaft 6,* Wiemer, K. & Kern, M. (Eds.), pp. 467-494, M.C.I.Baeza Verlag, ISBN 3881226338, 9783881226332, Witzenhausen.

Werner, W.; Scherer, H.W & Olfs, H.-W (1988). Influence of long-term application of sewage sludge and compost from garbage with sewage sludge on soil fertility criteria. *Journal of Agronomy and Crop Science*, Vol. 160, Issue 3, pp. 173–179.

Woods, W. I. (2003). Soils and sustainability in the prehistoric new world, In: *Exploitation Overexploitation in Societies Past and Present,* Benzing, B. & Hermann B. (Eds.), pp. 143–157, Lit Verlag, Münster.

Woods, W. I. & Mann, C. C. (2000). The good earth: Did people improve the Amazon basin? *Science* 287, 788.

Woolf, D.; Amonette, J. E.; Street-Perrott, F. A.; Lehmann, J. & Joseph, S. (2010). Sustainable biochar to mitigate global climate change. *Nature Communications* 1, 56.

Yoshizawa, S.; Tanaka, S.; Ohata, M.; Mineki, S.; Goto, S.; Fujioka, K.; Kokubun, T. (2005). Composting of food garbage and livestock waste containing biomass charcoal, In: Proceedings of the International Conference and Natural Resources and Environmental Management, 08.10.2011, Available from: www.geocities.jp/yasizato/Yoshizawa6.pdf

Yuan, J.-H.; Xu, R.-K.; Qian, W. & Wang, R.-H. (2011). Comparison of the ameliorating effects on an acidic ultisol between four crop straws and their biochars, *J Soil Sediment*, Vol. 11, pp. 741–750.

Zimmerman, A. R.; Gao, B. & Ahn, M.-Y. (2011). Positive and negative carbon mineralization priming effects among a variety of biochar-amended soils. *Soil Biology and Biochemistry* 43, 1169-1179.

Zinati, G.M.; Li, Y.C. & Bryan, H.H. (2001). Utilization of compost increases organic carbon and its humin, humic and fulvic acid fractions in calcareous soil. Compost Science & Utilization, 9: 156 - 162.